LESSON STUDY WITH MATHEMATICS AND SCIENCE PRESERVICE TEACHERS

This insightful volume offers an overview of the fundamentals of lesson study practice in US teacher education as well as examples from math and science teacher educators using lesson study in their local contexts.

The number of teacher educators using lesson study with preservice teachers is small but growing. This book is aimed at teacher educators who may want to try lesson study in university contexts without the challenge of translating the practice from the K-12 context on their own. In this volume, lesson study is broadly overviewed, attention is given to its constituent steps, and examples of lesson study in preservice contexts are shared. Given the broad array of teacher education program designs, numerous contingencies guide teacher educators in their implementation of lesson study, given their contextual affordances and limitations.

The lesson study descriptions and cases in this book will support teacher educators and scholars across subject specialties and geographic lines, as they seek instructional frameworks to advance their pedagogical goals.

Sharon Dotger is a professor of science education in the Syracuse University School of Education and the faculty director of teacher education and undergraduate studies. She teaches courses in elementary and secondary science methods and courses in curriculum, learning theory, teacher professional development, and science education research.

Gabriel Matney is a professor of mathematics education in the College of Education and Human Development at Bowling Green State University. He teaches mathematics education and methods courses for K-12 teaching programs.

Jennifer Heckathorn is an instructor in teacher education at Syracuse University. Her research focuses on teachers' decision-making about their professional development experiences. She has prior experience as a public school teacher and administrator.

Kelly Chandler-Olcott is Laura J. & L. Douglas Meredith Professor for Teaching Excellence and dean of the Syracuse University School of Education.

Miranda Fox is a secondary mathematics teacher in Ohio. She has a master's in curriculum and teaching with a computer technology endorsement from Bowling Green State University. In graduate school, Miranda facilitated lesson study and conducted research on PSTs experience with lesson study in Thailand.

WALS-Routledge Lesson Study Series

Series editors: Christine Kim-Eng Lee, Catherine Lewis, Kiyomi Akita and Keith Wood

This series aims to provide opportunities for researchers and practitioners in Lesson Study to share their work beyond the boundaries of their countries to an international audience. Lesson Study is increasingly popular as a tool for improving the quality of education and schools around the world. As many countries are adapting and contextualizing Japanese Lesson Study to their own needs in response to educational and curriculum reforms cognizant that what matters most is what happens in classrooms and its impact on teachers and students. As Lesson Study originates from Japan, there is also a need for English Language readers around the world to understand more deeply the underlying philosophies, policies, and practices of Japanese Lesson Study in the cultural contexts of their schools and classrooms. As well as original works in English from leading figures in Lesson Study, this series will also make available outstanding Lesson Study publications originally written in Japanese but extended and revised for an English audience.

Teacher Professional Learning through Lesson Study in Virtual and Hybrid Environments
Opportunities, Challenges, and Future Directions
Edited by Rongjin Huang, Nina Helgevold, Jean Lang, and Heng Jiang

Lesson Study with Mathematics and Science Preservice Teachers
Finding the Form
Edited by Sharon Dotger, Gabriel Matney, Jennifer Heckathorn, Kelly Chandler-Olcott and Miranda Fox

LESSON STUDY WITH MATHEMATICS AND SCIENCE PRESERVICE TEACHERS

Finding the Form

Edited by Sharon Dotger, Gabriel Matney, Jennifer Heckathorn, Kelly Chandler-Olcott, and Miranda Fox

Routledge
Taylor & Francis Group

LONDON AND NEW YORK

Designed cover image: Shutterstock

First published 2024
by Routledge
4 Park Square, Milton Park, Abingdon, Oxon OX14 4RN

and by Routledge
605 Third Avenue, New York, NY 10158

Routledge is an imprint of the Taylor & Francis Group, an informa business

British Library Cataloguing-in-Publication Data
A catalogue record for this book is available from the British Library

ISBN: 978-1-032-35344-9 (hbk)
ISBN: 978-1-032-35342-5 (pbk)
ISBN: 978-1-003-32643-4 (ebk)

DOI: 10.4324/9781003326434

Typeset in Times New Roman
by KnowledgeWorks Global Ltd.

This book is dedicated to the students and host teachers who have allowed us and our students to learn from, with, and alongside them.

CONTENTS

FIGURES AND TABLES

Figures

Tables

ACKNOWLEDGMENTS

A number of individuals and groups have contributed to the development of this book in ways that we need to acknowledge:

- Our students and host teachers, who have generously allowed us to learn from and with them;
- The advisory committee for the Lesson Study for Mathematics and Science Teacher Educators Conference (LSMSTEC): Catherine Lewis and Shelley Friedkin, who provided feedback on early iterations of the conference and the book proposal;
- All participants at LSMSTEC, who helped to stretch our imaginations about what lesson study in teacher education can look like;
- Jessica Belcher, events planner extraordinaire, who assisted us with the logistics and implementation of LSMSTEC;
- Mary Jo Hoeft, a graduate student at Syracuse University, who stepped in to LSMSTEC to lead a small group session;
- All of the authors of chapters in this book, who have kindly drafted and revised numerous times to help bring this book into existence;
- Martin Walls, Director of Communications and Marketing at Syracuse University School of Education, who formatted the figures in this book for publication;
- Our editors at Routledge, Katie Peace, who began the process with us, and Khin Thazin, who shepherded us through the end of the process;

This book is the result of the Lesson Study for Mathematics and Science Teacher Educators Conference (LSMSTEC), which was funded by the National Science Foundation–award #2010137.

Furthermore, individual chapters in this book report on projects that were funded by a variety of National Science Foundation awards. These include:

- National Science Foundation, Division of Undergraduate Education award #1930950 & #1930971 (Chapter 3);
- National Science Foundation award #1813127 (Chapter 14);
- National Science Foundation Robert Noyce Capacity Building Grant: Urban STEM Teacher Capacity Building, award #1540805 (Chapter 12);
- National Science Foundation Robert Noyce STEM Teacher Scholarships & Stipends grant: Noyce Urban STEM, award #1660506 (Chapter 12);
- National Science Foundation Noyce STEM Teacher Scholarships & Stipends Grant: A Community-Based Approach to STEM Teaching & Learning, award #2151027 (Chapter 12)

Any opinions, findings, and conclusions or recommendations expressed in this material are those of the authors and do not necessarily reflect the views of the National Science Foundation;

Finally, we would like to acknowledge and thank our families: Without your understanding, guidance, and grace, this book never would have happened.

PREFACE

This volume makes major new contributions to the understanding of preservice lesson study. Although focused heavily on STEM (science, technology, engineering, and mathematics) in North America, this volume has implications for teacher educators in every discipline and every region of the world and for lesson study outside of the preservice setting.

Deftly organized around a shared framework—the key tasks of the Study, Plan, Teach, and Reflect steps of the lesson study cycle—the chapters are closely integrated and highlight how much we can learn from examining each other's lesson study practice. The volume captures a rich landscape of preservice lesson study, adapted to diverse institutional and policy contexts; Figure 1.1 summarizes the landscape. If you are facing a design challenge in preservice lesson study—for example, integrating two STEM disciplines (Chapter 12), or creating a meaningful lesson study experience for aspiring teachers before they have taken education courses (Chapter 17) or if you are interested in the multiple roles teacher educators take on in lesson study (Chapter 3)—you are likely to find an example focused on that challenge.

I will mention just a few of the ideas I took away from reading this volume. First, the *prepare* step (before the lesson study cycle begins) lays important groundwork for the lesson study cycle. For example, Chapter 2 describes how preservice teachers study the TRU (Teaching for Robust Understanding) Framework prior to beginning the lesson study cycle, allowing them to build awareness of their own instructional decision-making before commencing the lesson study work. Chapter 4 explains the careful negotiation with field partners that lays a foundation for lesson study to support the goals of the mentor teacher and school leaders, as well as those of the preservice program.

Several chapters make the case that the *study* step of lesson study is particularly important for preservice educators, and you will find powerful resources and suggestions for strengthening this step. For example, Chapter 5 elaborates five dimensions of content study, enriching our ideas about what it means to study content; the chapter also highlights the power of having preservice teachers identify a personal learning goal. A set of chapters on equity-focused instruction suggest a range of useful resources, with concrete cases on how to study them and incorporate them throughout the cycle. These resources include the Rights of the Learner, three anti-racist pedagogies, and both dominant and critical approaches to equity. Chapters richly describe the role of content knowledge in supporting preservice teachers to enact equity-focused pedagogy (Chapter 9), practices that allow preservice teachers to build the trust needed to examine their pedagogy (Chapter 10), and how preservice teachers' own experience of the Rights of the Learner as preservice students supports them to enact this pedagogy in classrooms (Chapter 11). The preservice literacy and mathematics collaborations described in Chapter 15 reveal that prospective teachers can learn more than just specific content during the *study* step; they can develop a broader appreciation of the vital role of studying outside content and research on the topic to be taught.

Chapter 6 provides a detailed account of the *plan* step, with particular attention to the ways that educators outside Japan need to provide support for elements (such as unit structure centered on student sense-making) that are prominent in Japan but may be missing elsewhere. The chapter unpacks many key ideas needed to make the *plan* step succeed, such as a focus on how the students experience the mathematics (not on finding activities for student engagement) and on "interaction problems" that develop agency, ownership, and identity as students explain and critique their ideas.

Another interesting theme of this volume is the way that lesson study structures need to change as prospective teachers move through their preservice education. For example, Chapter 7 describes three effective protocols for the *teach* step that build on each other, for use over the course of teacher preparation, beginning in introductory courses and continuing in more advanced forms through upper-level methodology courses.

Chapter 14 explores three tools to deepen lesson analysis and examination of culturally responsive teaching during the *reflect* step, with explanation of how to support individual sense-making (before collaborative reflection) and study of the tools (early in the lesson study cycle) for maximum impact. Chapter 16 further deepens our understanding of the *reflect* step, by exploring the role of observational notes and collaborative peer reflection in supporting preservice teachers' learning.

Readers are likely to encounter enticing new adaptations of lesson study. A "Watch Party" for favorite videos from online lesson study (along with specifics of the GoReact platform to create lesson tags) caught my attention in Chapter 13, especially after reading about the changed attitudes toward lesson planning that preservice teachers reported and the ease of sharing the lessons with site-based mentor teachers as well as university instructors and classmates. And who wouldn't want to attend the 3-week summer program "Camp Questions" described in Chapter 8, where preservice teachers build the understanding of literacy instruction while local youth take part in an enrichment program?

Together, the chapters of this volume provide an exceptionally exciting vision of preservice lesson study as a place where all prospective teachers build dispositions, knowledge, and commitments needed to teach well. Thank you to the volume editors and authors for sharing their experiences and vision.

Catherine Lewis

INTRODUCTION

Tracing Lesson Study's Use with Preservice Teachers: From Origins to Present Day

Sharon Dotger, Gabriel Matney, Jennifer Heckathorn, Kelly Chandler-Olcott, and Miranda Fox

Teacher educators—professionals who prepare preservice teachers (PSTs)—are a hopeful and ambitious group. They imagine they can generate positive change in society by preparing teachers who are committed to equity, eager to learn, capable of listening to children in classrooms, and adjust their practice because of what they hear and learn. We, the five co-authors of this introductory chapter, are happy to include ourselves among this group. Yet, mixed with this happiness is an acceptance that teacher education can continue to improve and we view lesson study as one mechanism to support improvement. The authors of the chapters in this book have incorporated lesson study into their teaching because they have experienced its power to improve their own teaching and learning, and that of others.

In learning to lead lesson study, teacher educators in the United States (US) have drawn from examples within the literature to use lesson study with candidates in their education programs. However, the lesson study examples most prevalent since the turn of the century almost exclusively come from K-12 contexts. Given the broad array of teacher education program designs, numerous contingencies guide teacher educators in implementing lesson study in the best way possible. Therefore, this book has two primary aims: One is to overview the fundamentals of lesson study practice in US teacher education. The second is to illustrate various cases from math and science teacher educators using lesson study in their local contexts. The lesson study description and the cases can support teacher educators in the range of content areas addressed within K-12 schooling and beyond the US as they seek instructional frameworks to advance their pedagogical goals.

DOI: 10.4324/9781003326434-1

What Is Lesson Study?

Lesson study is a set of professional routines, carried out by teachers working collaboratively, and organized sequentially with an aim to test how instructional ideas impact student learning while live lessons unfold. Like design-based research, lesson study tests a research hypothesis articulated by the participants about a problem of instruction (Campanella & Penuel, 2021). The test is conducted within the classroom during the research lesson. Data about its impact on students is collected in real-time, used afterward to evaluate the lesson design, and drives further iterations of lesson study. Importantly, lesson study is not intended as a means to design a perfect lesson, but rather to investigate instructional routines that can be used in daily instruction and improved upon over time.

Other teacher educators have noted that classroom teachers rarely get the opportunity to practice, in the sense that once they are employed, they rarely get to rehearse, repeat, or experiment with variations in their instruction (Grossman et al., 2009). Lesson study provides teams of teachers an opportunity to practice, in that it slows down the essential skills of planning, rehearsal, evaluation of instructional materials and standards, studying student work, and taking student perspective during observation—key components of practice-based teacher education (Hammerness et al., 2020). These components of lesson study are connected, creating a coherent, focused experience that can catapult a change in teachers' instruction for future lessons. In this way, lesson study is practice for professional teaching.

Like other forms of practice, lesson study relies on a series of activities, often supported by a knowledgeable other (a person whose role is described in greater detail in later chapters). For this work, we draw from five sources that identified essential features of lesson study with practicing teachers (lessonresearch.net; Lewis et al., 2019; Takahashi & McDougal, 2016, 2019; Seleznyov, 2018). We used these articles to synthesize essential features of lesson study and compare how they described lesson study's central components, which we call steps throughout the book: *Prepare, study, plan, teach,* and *reflect.* Each of these steps can be broken down into finer-grained aspects—and chapters in the book will help you visualize these in greater detail. For the purpose of introduction, however, we overview them here.

Readers familiar with lesson study resources might be surprised to see *prepare* listed as one of the steps of lesson study. We found it on lessonresearch.net, a website updated periodically as its authors (Catherine Lewis, Shelly Friedkin, and colleagues) learn more about how to communicate with teachers and instructional leaders who are doing lesson study. In our view, recognizing *prepare* as a step provides an opportunity to formally acknowledge the work that needs to be done before the *study* step can be initiated, although we acknowledge that many lesson study scholars include this work within the *study* step. During the *prepare* step, teams organize, set agendas, and determine times for collaboration. They gather the resources they will use throughout the remaining cycle and set norms to guide their work.

From here, the team of teachers transition to the *study* step, where they study their standards, instructional materials, research about instruction and content, and think about equitable teaching and learning. They articulate a research theme and connect the theme to the ideas they develop while studying. This step guides them into the next step: *Plan*. During the *plan* step, they construct a detailed document that is part lesson plan, part record of their design rationale, and part a means to communicate to outside observers or other lesson study practitioners. Teachers may also rehearse the research lesson, which may include creating a mock-up of the design of the board or other public representations of student thinking, such as the documents they expect students to generate while learning.

Following *study*, the team is ready to enter the *teach* step. In this step, one team member delivers the lesson to students, while the other team members observe students for evidence of learning as the lesson unfolds. The observations gathered by the team are shared following the lesson and this begins the *reflect* step, where discussion of the evidence is used to evaluate the lesson's learning and connect the lesson design to future lessons the teachers will enact. Taken together, lesson study practitioners refer to the completion of these steps as a cycle of lesson study.

Where Did Lesson Study Come from?

In 1998, a publication entitled "A Lesson is Like a Swiftly Flowing River" by Lewis and Tsuchida re-introduced the world outside of Japan and China to lesson study. One year later, *The Teaching Gap* by Stigler and Hiebert, an analysis of data from the Third International Mathematics and Science Study, added additional examples for English speakers. Both attributed Japanese teachers' skills in teaching for students' understanding in science and mathematics to lesson study and offered initial descriptions of how Japanese teachers engaged in it. Over time, as interest in lesson study has grown, Japanese scholars have explained its development and prominence in Japan. This explanation is worth repeating here, in part because it may help teacher educators understand that the cultural divide between Eastern and Western teacher education may not be as wide as some may think.

Lesson study's roots trace to a collaboration between Edward Sheldon, a superintendent, teacher educator, and eventually president of the State University of New York (SUNY) system, and a Japanese teacher educator, Takamine Hideo, in 1872. While preparing elementary teachers in Oswego, New York, Sheldon taught his future teachers to design and teach object lessons through the use of demonstration lessons and criticism lessons. Object lessons are those where students were expected to learn via working, talking, and thinking with manipulatives. In demonstration lessons, teacher candidates at Oswego watched object lessons taught by expert teachers. To practice the object lesson method themselves, candidates would teach a lesson with their classmates and teacher educator in attendance as observers. When the lesson ended, everyone discussed what occurred. These were criticism lessons. While Sheldon did this, Japan was revamping its education system to broaden access to its citizens and therefore professionalize the teaching force. Takamine Hideo came to Oswego to learn these methods and deployed them at Tokyo's first Japanese normal school. Additionally, around the same time, Marion Scott, a teacher from the San Francisco area, traveled to Japan to teach demonstration lessons in new Japanese normal schools.

This history means that in Japan, lesson study began in teacher education and continues there today. It was also exported to public schools, where it thrives to the point that US researchers conducting separate projects in Japan in the 1990s were able to find it without explicitly looking for it (Lewis & Tsuschida, 1998; Stigler & Hiebert, 1999). Their reports helped others envision lesson study as a practice for in-service teachers. It has now been reimported by some teacher educators into their courses in their teacher education programs in the US, which will be described in greater detail in the chapters in this book. While there is little doubt that Japanese teachers and teacher educators refined Sheldon's original ideas (*q.v.,* Sheldon, 1870), lesson study's circuitous route back to US teacher education means that we are reviewing some of the wisdom of original ideas of preparing novice teachers, thus shining a light on the similarities between teacher preparation across countries.

How Did This Book Come to Be?

This book is an outcome of a national working conference, the Lesson Study for Mathematics and Teacher Educators Conference, held April 14 and 15, 2021, that included 33 teacher educators and the book's editorial team. Held online because of the COVID-19 pandemic, the conference included various sessions, which we will describe in greater detail below.

A grant from the National Science Foundation (NSF) supported the conference. The seed of that proposal was generated by Sharon, Gabriel, and Miranda as they discussed their own experiences and conversations with other colleagues. They realized that other US teacher educators were conducting lesson study within their teacher education programs. Yet, they suspected that more teacher educators were doing the work than they knew. Furthermore, given their respective foci as science and mathematics teacher educators, there were not many venues where the expertise of both content areas could come together to discuss lesson study practice. Sharon and Kelly, for example, had experience collaborating with other teacher educators to launch lesson study efforts in a program for master's students (Chandler-Olcott et al., 2019, 2021). Additionally, Sharon and Jennifer had collaborated on multiple lesson study projects, some with K-12 teachers and others with PSTs in local public schools (Dotger et al., 2021). Similarly, Gabriel had been working with different school districts around the US to help them establish lesson study as part of their professional learning process (Weaver et al., 2021). Gabriel and Miranda had been working to research lesson study in international PST education contexts (Matney & Fox, 2022). The team brought these numerous lesson study experiences and collaborations together, and an NSF proposal was generated.

Once the grant proposal was funded, Sharon and Jennifer met weekly to draft a conference agenda, coupled with a detailed plan for facilitating each session within the 2-day meeting. They shared their draft agendas with the remaining team and together, we refined a process to elicit participants' lesson study practices while also deepening the communities' shared understanding of lesson study. As the agenda was built, participants applied to the conference and once accepted, submitted a poster that described their current lesson study practice. From the participants' Lesson Study for Mathematics and Science Teacher Educators Conference (LSMSTEC) applications, we knew that participants had a variety of understandings about lesson study and its use in their own contexts. As a way to build shared understanding about lesson study and set a course for the conference, we selected an article as a shared reading that could be referenced throughout our activities. The article we selected to serve as the focal text was by Catherine Lewis and colleagues (2019). We chose this chapter because it was recent, defined and explained the steps of lesson study, and connected theoretical frameworks to practice-based goals.

During the conference, we facilitated small-group sessions where participants shared posters of their lesson study practices and discussed these practices with one another. The posters were archived in a shared folder for all conference participants to review. After this first sharing session, we led a whole-group discussion of the focal text, which included detailed descriptions of four of the lesson study steps as applied in K-12 contexts. We intentionally scheduled the sharing of practice before discussing the focal text, mainly because we anticipated that the focal text might be seen as a definitive text that might result in undue comparison against an individual's practice. We wanted the participants' lesson study practices to be brought forward for discussion, compared and contrasted with each other to be contextualized, and then further discussed as compared to a descriptor of lesson study in the in-service context. During that conversation, we leveraged a series of digital tools to capture participants' thinking at the moment, which served as an additional archive for authors to review as we constructed this book.

The next step of the working conference sent participants into one of four small groups. Each group was assigned a lesson study step *(study, plan, teach, reflect)* and tasked with synthesizing the ideas shared by colleagues in the poster session with the ideas discussed from the book chapter. Each group had a Google slide to create, which captured the patterns in their discussion and supported them in sharing their thinking with other groups. Once the slides were created, one speaker from each group shared their thinking, and other groups offered questions and suggestions for further work. These small groups worked together a second time to respond to the questions and adjust their work. These slides and the discussion notes that resulted from their presentations supported the articulation of the lesson study step chapters (numbers 5 to 8) later in this book.

Separately, we held a session that focused on *prepare*. In this session, participants reviewed the information on the website about preparing and then used a structured reflection document to generate their ideas about how they conducted similar work in their context. They shared their work with a partner and uploaded their revised document to a shared folder for all participants. This work informed Chapter 5—a chapter focused entirely on this important and often invisible lesson study step.

The final working session of the conference focused on generating ideas for the contents of this book. We began in Google slides, generating a list of ideas that the participants imagined could be included in a book that focused on lesson study in teacher education. We turned this list into a survey, asking participants to identify three possible ideas they were interested in contributing to the book, along with potential co-authors. Once all the surveys were complete, we sorted the results, matched co-authors, generated an initial book outline, and constructed writing invitations for each authoring team according to their preferences.

By this point in the conference, attendees were more confident a book was a reasonable outcome from this work because, as participants shared their knowledge and practices, they commented on how much they were learning about aspects of lesson study. For example, some participants stated that they were doing much of the work in *prepare* but not naming it as explicitly as they could be. Others noted they would place more emphasis on *study* in future work with PSTs, while some picked up new nuances to *plan* that they would modify in their future practice. Furthermore, a few found new ideas for including PSTs more fully in *teach*, and we discussed at length the role of *reflect* in lesson study specifically, as different from the notion of reflection in teacher education more broadly. With all these ideas emerging among conference participants, we found it reasonable that others might benefit from access to similar ideas and thus, we included them in this book.

What Will You Find in This Book?

Section I builds on the initial overview of lesson study history presented here, details its history and migration into teacher education, and demonstrates its value as a pedagogical tool. In Chapter 1, Jennifer Heckathorn and Sharon Dotger overview the teacher education landscape in the United States, describing its variability to contextualize teacher educators' decisions when deploying lesson study with PSTs. In Chapter 2, Nadia Kennedy and Jesse Wilcox provide two compelling examples of how they used complete lesson study cycles to study and improve PSTs' decision-making skills. Building from this general description, in Chapter 3, Gloriana González, Omar Hernández Rodríguez, and Wanda Villafañe Cepeda pivot to discuss the intertwined roles teacher educators play during the lesson study process, helping to contextualize the more detailed descriptions that readers will encounter in later chapters. These chapters are early in the book so readers can develop their understanding of how the steps fit together and be ready for the details about the steps presented in Section II.

Section II consists of five chapters dedicated to deepening the reader's understanding of each step of the lesson study cycle. Each chapter focuses on one step of the cycle, discussing its defining features and providing examples of how lesson study in PST education is different from its use with in-service teachers. We begin the section with Chapter 4, written by Jenifer Hummer and Kristin Lesseig, which identifies critical features of the *prepare* step. We include it here to highlight those things a teacher educator must consider and do before implementing lesson study, especially with field partners. Rachelle Rogers, Ryann Shelton, and Trena Wilkerson co-author Chapter 5, which looks at the *study* step. These authors help readers understand the importance of *study* within the cycle, as it can be overlooked if practitioners rush to lesson planning too quickly. Chapter 6 is dedicated to the *plan* step and is written by Kevin Reins and Matthew Melville. They discuss their practices for supporting PSTs in developing research lesson plans that build upon the prior lesson study steps. Chapter 7, written by Rosemarie Michaels and Nicole Glen, tackles the *teach* step, acknowledging the simultaneous teaching and observation of students that happens in this portion of lesson study, and describing various models for the *teach* step. Finally, in Chapter 8, Kelly Chandler-Olcott and Sharon Dotger discuss the *reflect* step of lesson study, explaining how to support PSTs in reflecting on the research lesson and how its evidence implicates their planning of future lessons.

In Section III, authors position lesson study to advance K-12 students' access to effective teachers, and therefore as a mechanism for increasing equity in education. In Chapter 9, Melissa Graham and Amy Roth McDuffie report on their project, which involved designing an experimental course and studying PSTs as they engaged in lesson studies with a research theme focused on equity. In Chapter 10, Curtis Taylor, Kristin Komatsubara, and Daisy Sharrock explain how they use lesson study to foster anti-racist pedagogies with their novice teachers. In Chapter 11, Crystal Kalinec-Craig, Keely Hulme, Karisma Morton, Colleen Eddy, Fardowsa Mahdi, Dittika Gupta, and Mark Montgomery report on a project they completed in three elementary math methods courses using Torres' Rights of the Learner framework to advance PSTs' awareness and development of equity-based practices.

Finally, Section IV builds on the learning in the previous sections to provide additional examples of how lesson study is used in PST education. In this section, the authors take care to explain how they have modified the traditional lesson study cycle—and perhaps individual steps—given variations in their contexts. The section begins with Chapter 12, written by Janelle Johnson, Nursen Konuk, and Mark Koester, which demonstrates how two in-service teachers used their learning through lesson study in their teacher preparation program, to build opportunities for interdisciplinary collaboration. In Chapter 13, Rita Hagevik and Irina Falls describe how to adapt lesson study from face-to-face to online environments. Gillian Roehrig and Jennifer Suh authored Chapter 14, which focuses on how video, coupled with structured mechanisms for noticing and supporting teacher learning, especially during the *reflect* step, can support teacher learning. Chapter 15, written by Hanna Haydar, Meral Kaya, and Joanna Weaver, provides examples of PSTs learning to use aspects of lesson study to integrate science and mathematics with literacy instruction. Joanna Weaver and Gabriel Matney wrote Chapter 16, which focuses on how to conduct lesson study with PSTs when traditional teaching placements in the field are not available. Finally, in Chapter 17, Christopher Nazelli and S. Asli Özgün-Koca describe how to overcome obstacles to full-cycle implementation by designing a lesson study experience early in PSTs' academic trajectory.

How Might This Book Be Relevant to You?

This book is grounded in its multiple influences. For example, its origins trace to the World Association of Lesson Study (WALS) itself, an international space to help teachers, teacher educators, and school leaders drive forward lesson study practice and research. Now well into its second decade, the WALS organization convenes annual international conferences to study the finer points of lesson study. Given the interests of lesson study around the world, we hope this book will be useful to international readers. The US is a large, diverse place with various contexts for teacher education and schools. We hope this diversity serves as an asset to connect with colleagues worldwide who are using lesson study and that it contributes to additional sharing of information about how lesson study is practiced in teacher education.

We anticipate that a large portion of the readership will be teacher educators. As such, teacher educators who want to try lesson study in their courses and programs will hopefully benefit from our attempts to translate the practice into our contexts. Teacher educators who wish to deepen and extend their lesson study practice will likely find new insights to help them and their colleagues. While most of the authors in this book are science and mathematics teacher educators, these subject area-foci contextualize, rather than foreground, the lesson study descriptions here. As a result, we hope that teacher educators from all content areas and grade levels will connect to this work.

This book can also be helpful to graduate students studying teacher learning or professional development. While the book focuses on lesson study specifically, readers will find multiple connections to themes in the teacher learning literature. This book might also help graduate students seeking to study lesson study in their dissertations, use lesson study in their classes, or collaborate with schools as part of other projects.

As some examples in this book illustrate, teachers in the schools who collaborate with teacher educators on lesson study projects are often critical to the projects' success. If such teachers read this book, they might deepen their understanding of the intent of lesson study in teacher education, enhance their collaborations with teacher educators to drive teacher and student learning, and contribute to the development of lesson study practices. They may also glean ideas they can implement with their school-based colleagues.

Deans and department chairs in schools and colleges of education may also find this book helpful. The book highlights the constraints and affordances for lesson study in teacher education, thereby assisting leaders in program design, field partnership arrangements, and more fully understanding teacher educators' practices.

Concluding Thoughts

Lesson study is a worldwide phenomenon of professional learning among teachers and teacher educators. It is a robust process that allows teachers and teacher educators an authentic look into their own practice, allowing them to work together to overcome the challenges of the profession. We hope this book helps others aspire to enact lesson study with PSTs to produce positive change in teacher education—and in education more broadly—both within and beyond the US context.

References

Campanella, M., & Penuel, W. R. (2021). Design-based research in educational settings motivations, crosscutting features, and considerations for design. In Z. Philippakos, A. Pellgrino, & E. Howell (Eds.), *Design-based research in education: Theory and education* (pp. 3–22). Guilford.

Chandler-Olcott, K., Dotger, S., Waymouth, H., Crosby, M., Lahr, M., Hinchman, K., Newvine, K., & Nieroda, J. (2019). Teacher candidates learn to enact curriculum in a partnership-sponsored literacy enrichment program for youth. *New Educator*, *14*(3), 192–211.

Chandler-Olcott, K., Dotger, S., Waymouth, H., Hinchman, K., & Newvine, K. (2021). Collaborative design to support digital literacies across the curriculum. In Z. Philippakos, A. Pellgrino, & E. Howell (Eds.), *Design-based research in education: Theory and education* (pp. 147–166). Guilford.

Dotger, S., Heckathorn, J., Whisher-Hehl, J., & Moquin, F. K. (2021). Providing high quality professional learning opportunities through a lesson study conference. *Innovations in Science Teacher Education, 6(4)*. https://innovations.theaste.org/providing-high-quality-professional-learning-opportunities-through-a-lesson-study-conference/.

Grossman, P., Hammerness, K., & McDonald, M. (2009). Redefining teaching, re-imagining teacher education. *Teachers and Teaching, 15*(2), 273–289. 10.1080/13540600902875340.

Hammerness, K., Klette, K., Jenset, I. S., & Canrinus, E. T. (2020). Opportunities to study, practice, and rehearse teaching in teacher preparation: An international perspective. *Teachers College Record, 122*(11), 1–46. https://doi.org/10.1177/016146812012201108

Lewis, C., Friedkin, S., Emerson, K., Henn, L., & Goldsmith, L. (2019). How does lesson study work? Toward a theory of lesson study process and impact. In *Theory and practice of lesson study in mathematics* (pp. 13–37). Springer.

Lewis, C. C., & Tsuchida, I. (1998). A lesson is like a swiftly flowing river: How research lessons improve Japanese education. *American Educator, 22*(4), 12–17.

Matney, G., & Fox, M.. (2022). Examining programmatic lesson study in preservice teacher education. In S. Bateiha & G. Cobbs (Eds.), *Proceedings of the 49th Annual Meeting of the Research Council on Mathematics Learning* (pp. 86–94). Grapevine, TX.

Seleznyov, S. (2018). Lesson study: An exploration of its translation beyond Japan. *International Journal for Lesson and Learning Studies, 7*(3), 217–229.

Sheldon, E. A. (1870). *A manual of elementary instruction: For the use of public and private schools and normal classes; Containing a graduated course of object lessons for training the senses and developing the faculties of children*. Charles Scribner & Co.

Stigler, J. W., & Hiebert, J. (1999). *The teaching gap: Best ideas from the world's teachers for improving education in the classroom*. Simon and Schuster.

Takahashi, A., & McDougal, T. (2016). Collaborative lesson research: Maximizing the impact of lesson study. *Zdm, 48*(4), 513–526.

Takahashi, A., & McDougal, T. (2019). Using school-wide collaborative lesson research to implement standards and improve student learning: Models and preliminary results. In *Theory and practice of lesson study in mathematics* (pp. 263–284). Springer.

Weaver, J. C., Matney, G., Goedde, A. M., Nadler, J. R., & Patterson, N. (2021). Digital tools to promote remote lesson study. *International Journal for Lesson & Learning Studies. 10*(2), 187–201.

SECTION I

Lesson Study and Preservice Teacher Education

1

CONTEXTUALIZING TEACHER EDUCATION AND LESSON STUDY IN THE UNITED STATES

Jennifer Heckathorn and Sharon Dotger

In many ways, teacher preparation in the United States resembles teacher preparation across the world: (mostly young) people interested in becoming teachers enroll in a college or university, take courses to build their knowledge and skills with content and pedagogy, and have opportunities to practice, receive feedback, and improve. For most teachers in the United States (almost 90%), that happens through traditional teacher preparation programs at Institutes of Higher Education (IHEs). However, other teachers enter the profession and study teaching as they practice. Furthermore, the United States is a complex conglomerate—where 50 individual states and 16 territories vary in governance. Taken together, teacher preparation in the United States, therefore, is an aggregate of programs, pathways, licensing, and credentialing bodies that provide numerous opportunities to induct new teachers into the profession.

In this chapter, we provide more context about teacher preparation in the United States as background for understanding differences that will be apparent in subsequent chapters. We then discuss how lesson study has been used in the United States and highlight its transition to teacher education, focusing on how that transition served as an impetus for the design of the Lesson Study for Mathematics and Science Teacher Educators Conference (LSMSTEC). Finally, we make recommendations for using this book to integrate lesson study in your teacher education context.

Teacher Preparation in the United States

In traditional teacher preparation, most preservice teachers (PSTs) study for 4 years at accredited, state-approved IHEs while living in two worlds: on campus, where they are students, and in K-12 classrooms, where they practice teaching. But there are other pathways through which individuals can be inducted into teaching. For

DOI: 10.4324/9781003326434-3

example, individuals who have completed a baccalaureate degree can enroll in accredited, state-approved graduate programs that lead to the same licensure as those who completed undergraduate programs. States also allow individuals to achieve teacher certification or licensure through alternate pathways, each with individualized components that recognize the multitude of experiences industry professionals-turned-teachers might bring to a classroom.[1]

Additionally, as researchers and practitioners struggle to identify the defining characteristics of effective teacher preparation, perhaps due to the lack of efficient data collection about teacher preparation program activities (Goldhaber & Ronfeldt, 2019), opportunities for other industries to edge their way into the preparation of teachers have arisen. Therefore, challenges to the cost, focus, effectiveness, structure, format, and ideological orientation of traditional teacher preparation programs have been mounted not just by philanthropic foundations and social entrepreneurs but by government entities and professional societies (Imig et al., 2011). They have formed, advertised, and implemented their own teacher preparation programs, which often look nothing like traditional programs.

Most future teachers in the United States still choose a traditional teacher preparation program. According to *Preparing and Credentialing the Nation's Teachers: The Secretary's 10th Report on Teacher Quality*, a report on the United States' preparation of teachers, the number of state-approved teacher preparation providers in 2015–2016 was 2,106 with a combined total number of preparation programs at over 26,000. Highlights of the report (which represents the most recent detailed accounting of teacher education practices in the United States for the school year 2012–2013) include a breakdown of the types of teacher preparation program providers: 70% of the teacher preparation programs were traditional programs, 20% were alternative programs based at IHEs, and 10% were alternative programs not based at IHEs. Altogether, almost 90% of all PSTs were enrolled in traditional teacher preparation programs.

For this reason, and because most of the authors in this book serve as teacher educators at traditional teacher preparation programs at IHEs in the United States, we turn our attention to describing how these programs are governed and operated.

Teacher Preparation Program Oversight

A major component of the teacher preparation program oversight in the United States is accreditation, a process through which IHEs meet acceptable levels of quality as set forth by either state or federal agencies or government-approved organizations (U.S. Department of Education [USDOE], 2016). The United States Department of Education formally recognizes several agencies deemed reliable authorities of quality education and training, two examples are the Association for Advancing Quality in Educator Preparation (AAQEP) and the Council for the Accreditation of Educator Preparation (CAEP). Accreditation requires the IHE to meet numerous evaluation criteria of the agency. Meanwhile, individual programs at each IHE are also approved by their own state's education department, making it so program graduates are eligible for state teacher certification. While many states maintain certification reciprocity agreements with other states, extending certification in one state to teachers in another, it is facilitated state-by-state, meaning that a teacher in Oregon might need additional requirements to teach in New York (and vice versa).

Teacher Preparation Program Pathways

With each teacher preparation program at every IHE evaluated on criteria that differ by both accreditor and state licensing boards, divergence in programming is bound to happen. These differences arise when each teacher preparation program sets guidelines for admission to the program and continue accumulating as PSTs move through their candidacy programs. Teacher preparation programs—and pathways within programs at the same IHE—are further delineated within each state by the content and grade levels PSTs will teach. Although most of these requirements consist of three main components—coursework, licensing exams, and field experiences—the pathways multiply in number, and the approaches to acquiring those components are numerous. In our university, for example, there are currently 17 different programs through which a student can become a certified teacher in New York State.

The differences across programs likely match the breadth of programs. For instance, each program sets the number and type of required courses to comply with their accrediting bodies, their IHE's requirements, and the state-governing boards. Even then, however, course syllabi, which individual course instructors typically set, can vary greatly, such that a history of education course at one university is not guaranteed to resemble a course with the same title at another institution.

One common element throughout teacher preparation programs, however, is coursework often referred to as "methods courses." In US methods courses, teacher educators help PSTs bring together their learning from their previous courses in content and teaching/learning. Methods courses teach PSTs how to teach students content, with a focus on a particular subject area. For instance, PSTs studying to become middle- and high-school social studies teachers learn instructional methodologies for engaging with the history, economics, and government content and the students they will teach while simultaneously studying the content's learning standards. Throughout the rest of this book, chapter authors refer to their own methods courses as that is a popular place for teacher educators to use lesson study with PSTs. In addition to offering an opportunity for PSTs to merge philosophy and practice, methods courses often have the added benefit of being connected to field placements.

In the United States, another feature of traditional teacher preparation programs is the opportunity for PSTs to practice their learning in field placements. Sometimes referred to as clinical practice, internships, or field experience, field placements occur when PSTs embed in local K-12 schools for a specific amount of time with specific goals. For instance, early in their program, PSTs might embed in a classroom 1 day per week for only 2–3 hours at a time to observe the teacher and students. Later, as they progress through their program, they enter another classroom for 3–4 hours, 3 days a week, to teach one or two lessons a week. Finally, their experience likely culminates with a "student teaching," where the candidates spend weeks in a classroom for the full school day and take over all teaching responsibilities within a discrete unit.

However, differences in expectations are also evident in the amount of time PSTs spend in classrooms during their teacher preparation programs. As an example, according to the National Council for Education Statistics, in 2012 (the last data available), Texas (20,828), New York (18,046), and California (11,080) were the states with the most completers of teacher preparation programs. However, their requirements for time spent in the field were different. In Texas, PSTs need 30 hours of field experience (15 of which could come from watching videos) before completing at least 70 days of student teaching. In New York, PSTs need 100 hours of field experience before completing 70 days of student teaching. In California, individual programs set their own requirements if PSTs amass 600 field hours. Assuming time spent in the school is 7 hours each day, that totals 520 hours in Texas, 490 hours in New York, and 600 hours in California. We recognize that hours spent in field placements is a crude measure of a program's effectiveness, especially when variety exists in how those hours are spent. Part of what has inspired the authors of the chapters of this book to use lesson study with their PSTs is the desire to use the field placement hours in ways that are likely to engage PSTs in meaningful learning.

As such, it is fair to say that two PSTs in different programs—while engaging in coursework on pedagogy and content, and field experiences where they put their learning into practice—can have substantially different cumulative experiences. Combine the variance in teacher preparation program design as dictated by the requirements of accrediting bodies and individual state certification requirements, with the variety of ways that IHEs can design their coursework and field experiences that provide distinctive experiences for their PSTs, and the conditions for differences in philosophical approaches, structure, opportunities, and coursework across teacher preparation programs abound. (And even, in some cases, within teacher preparation programs at the same IHE.) With this variance in mind, we turn now to describe the modifications that are made to lesson study when it is taken up in PST education.

Lesson Study in the United States

In the *Introduction*, we overviewed the history of lesson study—describing how it originated in teacher education in the United States, was taken up in Japanese teacher education, became a widespread practice for both in-service and PSTs there, and was exported back to the United States where it was taken back up with in-service teachers. There are a few US-based lesson study projects that have influenced our understanding of how lesson study can be conducted in the in-service setting (Lewis et al., 2019; Takahashi & McDougal, 2018). From this work, we have conducted lesson studies with in-service teachers investigating the relationships between science and art in the elementary setting (Dotger & Walsh, 2015), public research lesson conferences about K-12 ambitious science teaching (Dotger et al., 2021), argued for lesson study as a means to advance multicultural science teaching (Dotger & Burgess, 2022), and described how lesson study is useful for advancing disciplinary literacy (Chandler-Olcott & Dotger, 2022).

Lesson Study in Teacher Preparation

As teacher educators, we have used lesson study with our PSTs (Matney & Fox, 2022; Weaver et al., 2021) and adjusted it, relative to its use with in-service teachers. The major reason for this, in our view, is that the teacher preparation program context has different boundaries than the in-service context and, therefore, can sometimes limit the ability of teacher educators to conduct a full lesson study cycle with PSTs. For instance, the semester PSTs are enrolled with the teacher educator bounds the time for the cycle. Likewise, the relationship between the teacher educator and host teacher(s) can determine whether PSTs are welcomed into a classroom to observe the research lesson. As mentioned in the *Introduction*, the purpose of the grant and our working conference was to explore the degree to which our experiences were shared by other teacher educators.

Therefore, to prepare the grant proposal that funded the conference, we conducted a systematic literature review of peer-reviewed published articles that discussed the use of lesson study with PSTs. That work uncovered 96 articles we read and coded for features of lesson study practice. We then compared the features identified in the literature on PST lesson study with the characteristics of in-service lesson study identified by four core sources (Lewis et al., 2019; Seleznyov, 2018; Takahashi & McDougal, 2016; Takahashi & McDougal, 2019) to identify areas where more specificity about the use of lesson study with PSTs was needed. Although this broader book discusses lesson study with PSTs in the context of the United States, we did not limit the articles we reviewed to only those discussing practice in the United States because doing so would have seriously limited the number of articles available for review, possibly limiting the identification of the scope of modifications made to lesson study. Having the broader scope of practice as a background enabled us to better plan for a potential range of practices LSM-STEC participants may have used.

The literature review revealed places where teacher educators' practices using lesson study with PSTs diverged and converged, which led to us wondering what practice in the United States looked like. For instance, in our literature review, we found four main models of lesson study team membership. In the most prominent model, PSTs engaged in lesson study within a K-12 field placement. Additionally, most teams were composed of only PSTs (typically between two and six). However, other models had PSTs collaborating with a mentor teacher, facilitator, or multiple types of others—including university researchers, host teachers, and/or university supervisors. We also found models where team members included university professors conducting lesson study with their PSTs as the students. In the final model, PSTs were not part of the official lesson study team but rather participated as observers of the research lessons. Therefore, the variation in the make-up of lesson study teams seemed like a concept worth investigating with the LSMSTEC participants.

As a result of our literature review, we developed Table 1.1. In this table, we use the core features of the steps of lesson study use with in-service teachers (as defined by Lewis et al., 2019; Seleznyov, 2018; Takahashi & McDougal, 2016; Takahashi & McDougal, 2019) to document what we learned about lesson study use with PSTs, and areas in the literature where more specificity is needed. In addition to *study*, *plan*, *teach*, and *reflect*, we borrowed from our friends at The Lesson Study Group at Mills College (lessonresearch.net) to include a *prepare* step in which we describe the activities teacher educators undertake to generate the conditions necessary to pursue lesson study with PSTs—both in their classrooms and in K-12 school settings. We then used this table as a guiding feature in developing areas where we wanted to focus our investigation into our colleagues' practices during LSMSTEC.

Based on Table 1.1 and our experience with variability in teacher preparation programs in the United States, we conjectured we would see variance in LSMSTEC participants' practices. We believed this variability generally would be exemplified in how our teacher educator colleagues designed and implemented their lesson study practice with their PSTs. We hoped that by giving everyone a similar frame from which to discuss their lesson study practice, we could capture the breadth and depth of that variability, discover similarities in practice, and drive rich conversations about the nuances of our practice, all of which we could archive and use in the writing of this book.

The systematic literature review gave us a frame of reference to understand our fellow LSMSTEC participants' practices. For instance, we learned that teacher educators worldwide frequently cited the promotion of PST learning as their reason for using lesson study with PSTs. Then they mentioned subgoals, such as planning instruction, developing pedagogical content knowledge, improving observational capacity, evaluating or designing instructional materials, linking theory to practice, improving collaboration, and engaging in assessment. Through LSMSTEC, we wanted to understand which of these goals was most important to our fellow teacher educators—and how their practices varied based on their prioritized goals.

TABLE 1.1 Connections between Lesson Study Use in Preservice Teacher Education and with In-Service Teachers.

	In-service lesson study	*Initial teacher education lesson study*	*Areas needing more specificity*
Prepare	Establish team	Team membership can include PSTs and hosts	Team composition
	Build schedule and agenda		Timing of lesson study cycles
	Establish norms	Some programs offer training in lesson study	How trainers built skills; who is trained
	Share documents	Some lesson study manuals shared with hosts	Documents used, authorship
Study	Access teaching/learning plan		
	Research theme/purpose	Mentioned in some cases	Who determines the theme? How?
	Topic/Identify a focus		Who determines the focus? How?
	Kyouzai Kenkyuu	Some include assessment of students' ideas	What is included? Who is included?
	Knowledgeable other		Who is the knowledgeable other?
Plan	Unit plan		The unit the lessons are embedded in
	Develop research plan	PSTs are frequently the author of a lesson plan	Lesson or research plan?
	Written proposal		Are all these proposals written?
	Lesson planning		Lesson plan or research plan?
	Knowledgeable other		Who is the knowledgeable other?
	Mock-up lessons with revisions	Some PSTs practice lessons before enactment	How are these conducted? Who attends?
	Kyouzai Kenkyuu Logistics Focus data collection Confirm lesson responsibilities		

(*Continued*)

TABLE 1.1 (Continued)

	In-service lesson study	*Initial teacher education lesson study*	*Areas needing more specificity*
Teach	Pre-lesson discussion Teach live lesson	Some happen in schools, some microteaching	What are the outcome differences?
	Observe the live lesson	A few use observation forms, some lessons are taped	
	Discuss with knowledgeable other		Who is the knowledgeable other?
Reflect	Review observation data Post-lesson discussion		How are the videos used? Who participates in these discussions? Is the teacher educator the commentator or someone else? Who is the outside expert?
	Commentary Outside expertise Connect to others/share	Only one article specified this	
	Consolidate/summarize new learning Written reflection added to research plan	Written reflections are frequent	

As a result, when the LSMSTEC participants designed a virtual poster to present on their lesson study practice, they shared the following information:

- The goals they had for utilizing lesson study—specifically, what the teacher educators hoped their PSTs would learn from participating in lesson study.
- Their IHE context and the impact that had on their lesson study structure.
- Descriptions of what activities were done in the *study*, *plan*, *teach*, and *reflect* steps.
- Design details, including:
  - Essential elements—features of the design that were necessary for lesson study to work;
  - Affordances—aspects of the design that supported or amplified their ability to do lesson study;
  - Constraints—aspects of the design that placed limits on their ability to do lesson study;
  - Adjustments—changes the teacher educators made to their lesson study process as a result of evaluating their previous cycles; and
  - Timescale—the amount of time they spent in lesson study with PSTs, including the number of weeks per cycle and the length of research lessons.

The results of what participants shared during LSMSTEC provided clarity on some areas needing more specificity from Table 1.1 and generated questions in response to others. Throughout the chapters that follow, you will notice the authors explain the context of their programs and lesson study use, shedding light on the ways teacher educators modified lesson study to meet their needs. For example, through LSMSTEC, we learned about the various ways teacher educators approach selecting the research theme (a teaching hypothesis that tests the relationship between the instructional moves the teacher makes and evidence of student learning—see Chapters 5 and 6 for additional details)—which turns out to be quite different than what typically happens when lesson study is done with in-service teachers. For instance, when the lesson study cycle includes PSTs who are embedded in a field experience, the content is likely to be driven by the instructional calendar of the cooperating school, whereas when lesson study is done by in-service teachers, the teachers have more autonomy to align the theme with their curricular interests—no matter the time of year. This is different from lesson study cycles that utilize microteaching (q.v., McKnight, 1971), where PSTs are limited to teaching their classmates during a short timescale, and allows the teacher educator greater flexibility in selecting the content focus of the lesson study cycle.

In Japan, lesson study's development in the early 1900s occurred in public schools "attached" to normal schools. In this connected context, Japanese teacher educators—originally trained from public lessons taught by American teachers in these Japanese normal schools—learned how to conduct public object lessons and discuss them with their students (Makinae, 2019). However, by the time Lewis, Stigler, and Hiebert helped Western scholars take a new look at lesson study 100 years later, its practices among Japanese classroom teachers had shifted. Informed by decades of work with national standards, shared instructional materials, and more than a century of lesson study practices, knowledgeable others emerged. Takahashi and McDougal (2018) help us understand the work of the knowledgeable other in an in-service setting. They describe a system with two knowledgeable others who work with a team during lesson study. Both knowledgeable others have "extensive knowledge" of lesson study, lesson topics, courses of study, instructional materials, and student learning. The first person might bring examples from other lesson study cycles or groups, assist in the construction of the research proposal, or provide feedback on the proposal prior to teaching. The second, sometimes referred to as a final commentator, attends the research lesson, observes carefully, and offers final comments at the end of the post-lesson discussion.

In the United States, the system of lesson study is not yet ubiquitous. Therefore, teacher educators wishing to utilize lesson study must seek out opportunities to learn—as limited or as variable as they may be. For US teacher educators, opportunities to develop our knowledge about lesson study—so that we might become knowledgeable others ourselves—may be challenging to come by. For example, Sharon learned about lesson study by (1) attending conferences about math and science teacher learning that described lesson study; (2) attending open research lessons at the Japanese School in Greenwich, CT, and at the Chicago Lesson Study conference; (3) attending research lessons at World Association of Lesson Study and the American Educational Research Association conferences; and (4) leading lesson study groups with public school teachers and opening them to outside observers over time.

Because there are few systematic processes for all teacher educators in the United States to engage in lesson study, there remains work to do to better specify the role of knowledgeable others in teacher education. In the chapters that follow, readers will notice that the nuances of the work of knowledgeable others in lesson study with PSTs continue to develop. In some cases, the teacher educator acts as both knowledgeable others while in others, the teacher educator invites another expert to work in the role of knowledgeable other with their PSTs. We view the role of the knowledgeable other in PST lesson study as a line of inquiry that can extend beyond this book.

Concluding Thoughts

As you read the following chapters, we encourage you to pay particular attention to how *prepare, study, plan, teach*, and *reflect* are defined and described by our colleagues and how those conceptualizations meet the varying needs of individual teacher preparation programs. In doing so, you might notice areas where the practices of the teacher educators represented in this book are similar—for instance, with our tendency to use our methods courses as the foundation from which we pursue lesson study—and where they are different—for instance, with the variety of products that we ask our PSTs to create across the steps of lesson study. Finally, you may be thinking about how your practice mirrors—or does not—what you have learned about teacher preparation context in the United States—and how you can take the innovative ideas of our colleagues and apply them in your situation. Hopefully, the examples in the following chapters will help you compare the varied benefits of the multiple models presented and think about how to structure your lesson study cycles based on your specific context.

Note

1 As of January 2023, some states have loosened their education requirements for teachers. For example, Florida now allows veterans to teach without obtaining certification, Oklahoma is moving to allow high school graduates the ability to be hired as teachers, and Arizona permits college students to work as teachers.

References

Chandler-Olcott, K., & Dotger, S. (2022). *Lesson study to support disciplinary literacy in middle school science* [Paper Presentation]. Paper presented to the 2022 Literacy Research Association Annual Meeting. Phoenix, Az.

Dotger, S., & Burgess, T. (2022). Lesson study – a multi-faceted approach to improving multicultural science teaching and learning. *International Handbook on Multicultural Science Education*. https://doi.org/10.1007/978-3-030-37743-4_18-2

Dotger, S., Whisher-Hehl, J., Heckathorn, J., & Moquin, F. K. (2021). Providing high quality professional learning opportunities through a lesson study conference. *Innovations in Science Teacher Education, 6(4)*. https://innovations.theaste.org/providing-high-quality-professional-learning-opportunities-through-a-lesson-study-conference/.

Dotger, S., & Walsh, D. (2015). Elementary art & science: Observational drawing in lesson study. *International Journal for Lesson and Learning Studies, 4*(1), 26–38.

Goldhaber, D. (2019). Evidence-based teacher preparation: Policy context and what we know. *Journal of Teacher Education, 70*(2), 90–101.

Imig, D., Wiseman, D., & Imig, S. (2011). Teacher education in the United States of America, 2011. *Journal of Education for Teaching, 37*(4), 399–408.

Lewis, C. C., Schaps, E., & Watson, M. (1995). Beyond the pendulum: Creating challenging and caring schools. *Phi Delta Kappan, 76*(7), 547.

Lewis, C., Friedkin, S., Emerson, K., Henn, L., & Goldsmith, L. (2019). How does lesson study work? Toward a theory of lesson study process and impact. In R. Huang, A. Takahashi, and J. P. da Ponte (Eds.), *Theory and practice of lesson study in mathematics* (pp. 13–37). Springer.

Makinae, N. (2019). The origin and development of lesson study in Japan. In R. Huang, A. Takahashi, and J. P. da Ponte (Eds.), Theory and practice of lesson study in mathematics: an international perspective (pp. 169–181). Springer.

Matney, G., & Fox, M. (2022). Examining Programmatic Lesson Study in Preservice Teacher Education. Proceedings of the 49th Annual Meeting of the Research Council on Mathematics Learning.

McKnight, P. C. (1971). Microteaching in teacher training. *Research in Education, 6*(1), 24–38.

Seleznyov, S. (2018). Lesson study: An exploration of its translation beyond Japan. *International Journal for Lesson and Learning Studies, 7*(3), 217–229.

Takahashi, A., & McDougal, T. (2016). Collaborative lesson research: Maximizing the impact of lesson study. *Zdm, 48*(4), 513–526.

Takahashi, A., & McDougal, T. (2018). Collaborative lesson research (CLR). In M. Quaresma, C. Winslow, S. Clivaz, J. P. da Ponte, A. N. Shuilleabhain, A. Takahashi (Eds.), *Mathematics lesson study around the world: Theoretical and methodological issues* (pp. 143–152). Springer.

Takahashi, A., & McDougal, T. (2019). Using school-wide collaborative lesson research to implement standards and improve student learning: Models and preliminary results. In R. Huang, A. Takahashi, and J. P. da Ponte (Eds.), *Theory and practice of lesson study in mathematics* (pp. 263–284). Springer.

U.S. Department of Education. (2016). *Office of postsecondary education, preparing and credentialing the nation's teachers: The secretary's 10th report on teacher quality*. U.S. Department of Education.

Weaver, J. C., Matney, G., Goedde, A. M., Nadler, J. R., & Patterson, N. (2021). Digital tools to promote remote lesson study. *International Journal for Lesson & Learning Studies, 10*(2), 187–201.

2

PRESERVICE MATHEMATICS TEACHERS' DECISION-MAKING DURING LESSON STUDY

Nadia Stoyanova Kennedy and Jesse Wilcox

Teaching is a complex task that requires hundreds of non-trivial decisions be made each day—all of which greatly impact student learning (Clough et al., 2009). Teacher education programs face the challenge of helping preservice teachers (PSTs) understand the complexities of teaching practice, and of facilitating their learning to teach in the dynamic environment of the classroom while not overwhelming them. To reduce these complexities for PSTs, teacher educators often teach lesson planning as a linear process, which does not reflect the complex realities of the classroom (Superfine, 2009). Instead, lesson study can be utilized to collectively study teaching practice, preserve its complexity, and facilitate PSTs' deeper, more sophisticated understandings of and consequent change of beliefs about teaching. Using lesson study, teacher education programs can scaffold PSTs' learning by using a cohesive research-based framework that provides a clear vision of the planning, practices, and decisions that effective teachers make (Darling-Hammond, 2006). This chapter presents two cases that describe the integration of teacher decision-making frameworks (DMFs) into lesson study and explores the role of these frameworks in the preparation of PSTs. More specifically, it seeks to provide a context that encourages PSTs to become aware of the complexities of teaching and to improve their pedagogical decision-making.

Teacher Decision-making

Clough et al. (2009) argue that a research-based framework can help PSTs "to understand crucial teaching decisions, and how those decisions interact to affect student learning" (p. 821). Such a framework can act as an "advanced organizer" (Ausubel, 1964)—a broad, general structure—to help PSTs plan and enact lessons that account for the complexities of teaching and learning. In the early 1970s,

DOI: 10.4324/9781003326434-4

several scholars started to focus on the importance of teachers' decision-making, conceptualizing it as developing an essential link between theory and practice, and utilizing decision theory as a way of trying to understand and improve teaching (e.g., Bishop & Whitfield, 1972). Bishop and Whitfield (1972) were among the first researchers to study teacher decision-making, deliberative discussions, and regular reflection in enhancing the quality of teaching. They distinguished between pre-lesson and within-lesson decisions as well as short- and long-term decisions. Pre-lesson decisions included determining objectives, content, method, and materials. Within-lesson decisions included such aspects as implementation and/or modification of pre-lesson decisions, language use, task selection and modifications, error correction, motivating students, and group dynamics. According to Bishop and Whitfield (1972), teachers base their decisions on prior experience, training, and practice. Each teacher develops an individual decision-making schema for making those decisions. Teachers' schema, they argue, link classroom situations to prior experience, values, and teaching goals, which guide decisions and consequent action. Their work recognizes the importance of teaching contexts, teacher knowledge, values, and goals in shaping decisions and teaching acts.

Since then, educational researchers have made major advances in developing a theoretical framework that describes what teachers need to know to teach effectively, as well as how that knowledge is manifested in classroom practice (e.g., Ball & Bass, 2000). For example, Alan Schoenfeld's (2012) theory of teaching-in-context links teachers' knowledge, goals, and orientations (including beliefs, values, preferences, etc.) to their decisions. Similarly, Clough et al. (2009) propose a DMF to help teachers understand crucial teaching decisions based on education research, and how those decisions impact learners.

In this chapter, we explore the utilization of two DMFs (Figure 2.1) with PSTs in the context of a modified lesson study format. We discuss the lesson study process for PSTs that is used in methods courses and explore its potential for informing their decision-making related to lesson planning and enactment. We examine two different cases of methods of teaching mathematics courses, each of which employed a different research-based framework to guide PSTs' decision-making. Each was designed to help PSTs conceptualize and reflect on the multiple decisions they make, and to understand their importance for student learning.

Case Study 1

Context

This case describes lesson study practice in a methods course for preservice mathematics teachers (PSTs), offered in an undergraduate teacher program that prepares teacher candidates to teach grades 7–12. The program is hosted by one of the colleges in a large urban university in the northeast United States. There were ten students in the methods course, all of whom were in their third or fourth year and had

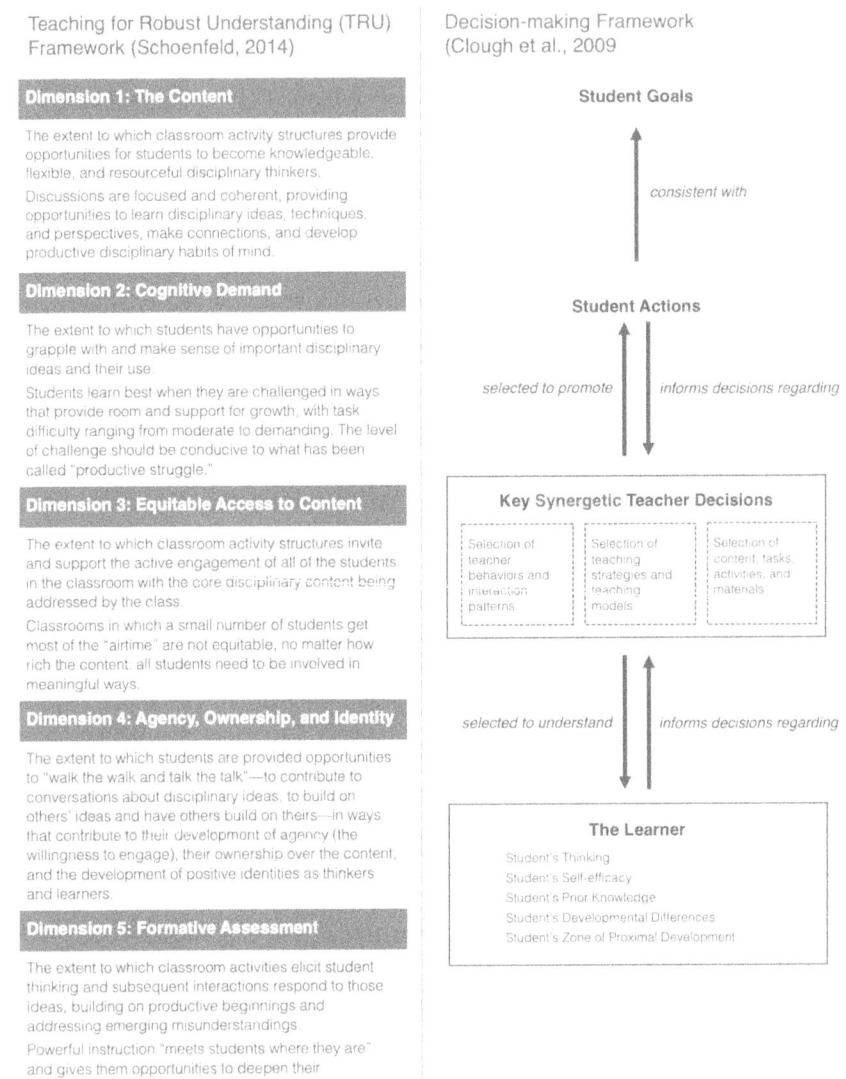

FIGURE 2.1   Schoenfeld's TRU and Clough et al.'s Decision-making Frameworks.

completed most of their required mathematics content and education courses. The class met once a week for 2.5 hours for 15 weeks. One of the goals of this course was to engage PSTs in the planning and rehearsal of teaching routines and enacting lessons. However, before PSTs can "enact" such routines, they need to be able to discern the basic elements of teaching practice. For the purposes of the course, we watched videos of classroom practice, which acted to enhance teachers' reflective "noticing" (van Es & Sherin, 2008), and to help them develop the capacity to reflect on classroom practice—that is, in identifying, analyzing, and interpreting aspects of teaching and evidence about student learning, and in formulating effective teaching strategies in response. For the purposes of identifying and analyzing essential elements of effective and powerful mathematics lessons and teaching, we used Alan Schoenfeld's *Teaching for Robust Understanding (TRU)* framework (2014). The framework identifies five elements of powerful mathematics classrooms: The content of a lesson; the cognitive demand determining its productive challenge; equitable access to content; the opportunities for students to develop agency, authority, and identity; and the use of assessment.

The Lesson Study

The lesson study was organized following the sequence of steps: *Prepare, study, plan, teach,* and *reflect.* During the *prepare* step, which took three weeks, we used videos of episodes of exemplary teacher-student lesson discussions as a means of inviting PSTs to analyze the five dimensions of the observed mathematical classrooms and lessons by using Schoenfeld's rubric (2014). During this step, we introduced a version of the lesson study plan template, which identifies four instructional columns: Learning tasks activities and key questions, anticipated student reactions and responses, teacher support (e.g., teacher responses to student reactions), and assessment (Lewis, 2002). When using this lesson plan template, PSTs must anticipate student reactions and identify forms of support that might mitigate expected student misconceptions and difficulties. We discussed several examples of lesson plans using this template. The entire class also used the lesson template to collectively plan a lesson introducing the concept of slope. The class decided collectively on learning goals, relevant standards, and possible learning activities, formats to facilitate them (e.g., independent work, small group, or whole class), questions that the teacher might ask to facilitate conceptual understanding, anticipated student responses, possible teacher responses in case of misconceptions, student questions, errors and feedback, various forms of assessment, and homework assignments. Lesson planning decisions were analyzed in reference to the TRU framework, followed by discussions about the affordances and the constraints that the specific planning decisions implied. The class also conducted a teaching mock-up during which the instructor taught selected portions of the lesson to the PSTs, who took the role of students. This allowed them to experience the lesson from the student's point of view, and to see how the instructor used the specific questions and teaching moves from the lesson plan, as well as to reflect on aspects of the lesson delivery in relation to the TRU framework.

After these initial discussions, the PSTs were ready to take part in the *study* step and engage with studying standards and content materials for lesson planning. We split the PSTs into five pairs, each of which worked as a team through the lesson study process. During the *study* step, each team identified a lesson topic of interest and engaged in studying relevant standards, curricular materials, and activities, all of which took two weeks. Each team also participated in consultations with the course instructor, who suggested additional resources or activities they might consider.

Next, the teams participated in the *plan* step, which took another two weeks. Each team identified the learning goals of the lesson in relation to standards, personal vision, preferences, and values. The teams used the -column lesson plan template, which had already been introduced. While working on lesson planning, we asked the teams to consult the TRU framework and rubric in making decisions pertaining to lesson design and delivery. Finally, we offered written feedback on each individual team's lesson plan and suggested revisions. After these were completed, each team met with the instructor to discuss lesson delivery and teaching strategies, and to select one or two of the lesson activities to be rehearsed by one of the team members. Due to time limitations, a portion of the lesson was selected for rehearsal rather the entire lesson.

Next in the *plan step*, we began with the lesson rehearsals. Each team needed at least two weeks to complete these. During week one, one of the team members rehearsed the selected lesson portion with all PSTs in the class as if they were her students. Each rehearsal lasted about 30 minutes and was videorecorded. Upon its completion, the entire class engaged in the *reflect* step. A copy of the lesson plan was distributed to all PSTs teachers. They discussed the plan and delivery using their observations of the lesson, following an established community reflection protocol. The protocol stipulated that the reflection starts with the pre-service teacher who delivered the lesson, after which the participant observers delivered positive feedback, followed by criticism and suggestions for improvement. The reflections were guided by the dimensions of the TRU framework. In accordance with TRU, they addressed questions such as: What did we find out about students' current thinking? What opportunities did students have to see themselves and each other as powerful thinkers and learners? How might we create more opportunities for each student to participate meaningfully? Next, within a week, the PSTs who had delivered the lesson watched and reflected individually on the video recording of her lesson delivery, and then discussed it with the instructor.

As a next task in the *reflect* step, and based on feedback and suggestions, the team met again to make revisions and improvements in their lesson plan. Then the rehearsal and reflect cycle just described was repeated with the second member of the team. Each team planned, rehearsed, reflected, and revised their lesson plan in this manner. Later the teams were also asked to consider ways to make provisions in the lesson design that would support English Language Learners (ELL) and students with disabilities. The repeated rehearsals and reflections took five weeks altogether. Each week, PSTs from two different teams delivered their selected lesson activities, followed by collective reflection. Finally, during the last three weeks of the semester, each PST engaged in the individual design of a unit plan, but still consulted with their other team member on decisions concerning design of activities, format, and lesson adaptations for ELL and students with disabilities.

Results of Lesson Study with the TRU Framework

PSTs reflections testified to the importance of the initial three weeks of viewing and analyzing video recordings of teaching. These activities helped PTSs notice aspects of the classroom environment they had not been aware of before, how the class discussions were facilitated, what kinds of questions the teacher asked, and what sorts of questions drove the development of ideas and conceptual development forward. Data from individual and focus group interviews showed that the reflections guided by the TRU framework also helped PTSs pay attention to how equitable a classroom was; how much students were encouraged to participate, and how student agency was promoted. In addition, at the completion of the lesson study cycle, we found noticeable differences between the initial lesson plans developed by the teams and their revised versions, as well as in the rehearsals of the two team members: Typically, the second team member used activities with improved design and was better able to launch and facilitate them. The revised lesson plans included tasks that were better sequenced, provided higher cognitive demand, better addressed the learning objectives, and made better use of small-group work and whole-class discussions. In the second cycle of rehearsals, the PSTs were generally more sensitive and responsive to student interventions, and made better use of questioning, whole-class discussion, and informal assessment.

Case Study 2

Context

The second case study describes lesson study practice in a preservice elementary program at a small liberal arts school in the Midwest. Eighteen PSTs took a science methods course in the fall and a mathematics methods course in the spring along with a practicum experience each semester. Both methods' courses were taught by the same professor. Students also took a literacy methods course in the fall and a social studies methods course in the spring taught by different professors. The methods courses met for 90 minutes twice a week for the first six weeks. For the remaining eight weeks, PSTs were in class 1 day a week and in the practicum experience another day of the week. The program concluded with a full-semester student teaching experience that students often took a semester or two after the methods sequence.

Throughout the elementary science and mathematics methods courses, we utilized Clough et al.'s (2009) DMF (Figure 2.1) as a visual representation of the synergistic nature of effective teacher decision-making. The DMF is not meant to capture all decisions a teacher makes, but instead serves as a tool for planning and diagnosing teacher decisions. This framework helped teachers consider how the decisions they make interact together to create the learning environment that their students experience (Clough et al., 2009). At the top of the DMF, is "student goals." PSTs developed between 10 and 15 student goals during the first week of class, which often included goals such as "students will be critical thinkers" and "students will communicate effectively." At the bottom of the DMF is "The Learner." This considers K-12 students' cognitive development, prior knowledge, social environment, self-efficacy, and affective factors. These two categories serve to help teachers determine the effectiveness of their teaching decisions, which is the central part of the framework that includes decisions about teacher interactions, teaching strategies and models, and the selection of content, materials, and tasks (see Figure 2.1). The DMF was used during lesson study to help PSTs understand the complexities of teaching and learning mathematics.

The Lesson Study

Just as in Case Study 1, the mathematics methods course followed the *prepare, study, plan, teach,* and *reflect* steps (Perry & Lewis, 2009). PSTs engaged in four lesson study experiences equally spaced throughout the semester including one in their practicum. Each lesson study was connected to an aspect of the DMF and included:

1 Teaching models (e.g., Launch-Explore-Summarize, Inquiry, Problem-based Learning);
2 Teacher interactions and mathematical communication (e.g., teacher questioning and students' written, oral, visual, and electronic communication);
3 Content and Standards for Mathematical Practice; and
4 Practicum lesson study.

At the beginning of the lesson study, we engaged PSTs in the *prepare* step through mathematics activities, critical incidents, and viewing video clips of lessons. Next, in the *study* step, we engaged PSTs, through discussions and readings, to make sense of their experiences in the *prepare* step. For example, when studying teaching models such as Launch-Explore-Summarize, we started the *prepare* step by investigating which lake of two large lakes in Okoboji, Iowa, is bigger (Switzer, 2015). PSTs explored how to measure the lakes on a map and, in the end, came to the conclusion one lake has a larger area and the other has a larger perimeter, which connects to standards 3.MD.C.6, 3.MD.C.7, and SMP 5 (National Governors Association, 2010). As we began the *study* step, we guided students to consider the structure of the lesson they experienced and then formally introduced them to Launch, Explore, and Summarize (Lappan et al., 2006). We then connected this to the learner and student goals aspects of the DMF by asking questions such as: "How did this lesson align with what we know about learners and learning theory?" "What student goals were promoted in this activity?" and "How could we better align the activity to promote more student goals and student learning?" After we introduced different teaching models, PSTs studied them in detail through class discussions, video clips of lessons, and readings. Importantly, we worked to model effective teaching and provided support for students throughout the *prepare* and *study* steps.

In the *plan* step, PSTs created lessons in pairs that aligned with a Common Core Mathematics standard and demonstrated their use of a particular pedagogical concept aligned with the DMF. For example, during the teaching model unit of mathematical methods, PSTs selected a teaching model discussed in class to use as a structure to plan their lessons. As PSTs planned their lessons, we walked around the room, asked questions, and prompted students to think about crucial decisions they need to make when planning and teaching.

In the *teach* step, PSTs taught a lesson with a partner while the remaining PSTs acted as elementary students. The *teach* step typically lasted two class sessions. Lessons were 30 minutes long and recorded. At the end of the *teach* step, the whole class engaged in a group reflection, which was the beginning of the *reflect* step. After all PSTs completed the *teach* step, PSTs engaged in the *reflect* step more deeply by writing a reflection on their lesson, as informed by the group reflection and our feedback. For example, with the teaching model lesson study, we asked students to reflect on written questions such as, "Why do you think the teaching model you selected was appropriate for this lesson?" "How did the teaching model promote student learning and student goals?", and "What went well in your lesson? What would you change and why?"

In summary, each lesson study took between 3–4 weeks to complete. The methods course had no lectures and required no rote memory or exams. Instead, in the *study* step, we modeled effective elementary mathematics activities and used those experiences, along with discussions, student questions, and readings to help PSTs conceptualize effective mathematics instruction. Students then worked to enact these pedagogical concepts through the *plan* and *teach* steps. Finally, in the *reflect* step, students analyzed their teaching by watching their teaching video and reflected on how to improve their practice.

Results of Lesson Study with the Decision-making Framework

To evaluate the *plan* step of lesson study, we analyzed PSTs lesson plans, recordings of their group work when they planned lessons, and interview data from three lesson studies (Lesson Study A, Lesson Study B, and Lesson Study C) that were completed. These data indicate that in Lesson Study C, PSTs were much faster at developing engaging lessons aligned to a standard, more focused on students' learning and engagement, considered differentiation and management decisions to a higher degree, and better anticipated students' responses. The *teach* step of the lesson study, analyzed through videos, also demonstrated many improvements. For example, PSTs asked more thoughtful, open-ended questions, scaffolded students thinking to a higher degree, and were better able to "think on their feet."

To investigate PSTs' reflections, we used the layered reflective model (Manouchehri, 2002) to determine the depth of PSTs' reflections, the extent to which they reflected on their decisions, and their judgments on how to improve them. This reflective model identifies five levels of reflection, organized from least to most reflection depth, and includes: Describing, explaining, theorizing, confronting, and restructuring. As the PSTs engaged in more lesson study cycles, PSTs more consistently reflected on deeper issues such as how PSTs could improve their teacher decision-making. For example, PSTs' reflections in Lesson Study A were pretty evenly distributed across all five categories, whereas in Lesson Study C, all participants were engaged exclusively in confronting and restructuring. The increasing depth of PSTs reflections, alongside specific references to the DMF, indicated that PSTs were likely internalizing and reflecting upon crucial teacher decisions outlined by the DMF.

Concluding Thoughts

In reflecting on the modified lesson study cycles that PSTs completed at both sites, we can draw some conclusions about PSTs' growth throughout the methods courses. First, PSTs' lesson plans at the end of the course better utilized small and whole-class discussions to guide students toward the learning objectives in question. Later lesson plans included lists of thought-provoking teacher questions, anticipated student responses, possible misconceptions and errors, and identified forms of teacher support that could address them. Second, PSTs' teaching practices improved. In both sites, they increased the level of student-to-student dialog and were better able to engage the whole class in activities and discussions. Third, in analyzing PSTs' reflections, we noticed that their writing changed to reflect the complexities of teaching and learning as the semester progressed. For example, we noticed many more PSTs wrote about the value of engaging students and working with their ideas, and the central importance of adapting lessons in relation to students' knowledge and learning needs. Finally, PSTs demonstrated a deeper understanding of the framework utilized, and an ability to connect those ideas to practice.

In this chapter, we described two different frameworks—Teaching for Robust Understanding (Schoenfeld, 2014) and the DMF (Clough et al., 2009). Both frameworks use student learning as a springboard to make effective teaching decisions. Although other effective frameworks exist (e.g., Windschitl et al., 2020), any framework chosen should be based on research and should keep the complexities of teaching intact. In addition, in order to integrate a DMF into the course, we recommend that it be used initially to analyze video lessons and lesson plans. We also recommend that PSTs be asked questions that make explicit connections to the framework. Questions such as, "How does this lesson offer high-cognitive demand activities?", "How could I have ruined this activity?", and "How does this lesson promote student agency?" can help PSTs unpack crucial teaching decisions.

Methods courses should strike a careful balance between theory and practice. Lesson study that utilizes a DMF can facilitate a synergy between the two and can help PSTs not only to understand mathematics teaching in its complexity, but to learn to deliver effective lessons. The DMF can support ongoing dialog about dimensions of lesson design and teaching that are often less visible, such as student authority, agency, and identity. It can also support PSTs in keeping important dimensions in mind as they plan and make key pedagogical decisions. Teaching in the lesson study provides a mechanism for PSTs to enact and reflect upon those decisions in a systematic way. As such, we suggest that the lesson study process in mathematics methods courses holds great promise for enhancing PSTs' reflective capacities, for helping them build the awareness necessary to make key synergistic decisions in the planning and enactment of teaching, and for furthering the overall improvement and pedagogical refinement of their classroom practice.

References

Ausubel, D. P. (1964). The transition from concrete to abstract cognitive functioning: Theoretical issues and implications for education. *Journal of Research in Science Teaching*, *2*(3), 261–266.

Ball, D. L., & Bass, H. (2000). Interweaving content and pedagogy in teaching and learning to teach: Knowing and using mathematics. In J. Boaler (Ed.), *Multiple perspectives on mathematics of teaching and learning* (pp. 83–104). Ablex Publishing.

Bishop, A. J., & Whitfield, R. C. (1972). *Situations in teaching*. McGraw-Hill.

Clough, M. P., Berg, C. A., & Olson, J. K. (2009). Promoting effective science teacher education and science teaching: A framework for teacher decision-making. *International Journal of Science and Mathematics Education*, *7*(4), 821–847.

Darling-Hammond, L. (2006). *Powerful teacher education: Lessons from exemplary programs*. Jossey-Bass.

Lappan, G., Fey, J. T., Fitzgerald, W. M., Friel, S. N., & Phillips, E. D. (2006). *Connected mathematics 2*. Pearson Prentice Hall.

Lewis, C. (2002). *Lesson study: A handbook of teacher-led instructional change.* Research for Better Schools.

Manouchehri, A. (2002). Developing teaching knowledge through peer discourse. *Teaching and Teacher Education*, *18*(6), 715–737.

National Governors Association. (2010). Common core state standards. *Washington, DC*.

Perry, R., & Lewis, C. (2009). What is successful adaptation of lesson study in the US? *Journal of Educational Change*, *10*(4), 365–391.

Schoenfeld, A. (2012). How we think: A theory of human decision-making, with a focus on teaching. In S. J. Cho (Ed.), *The proceedings of the 12th international congress on mathematical education* (pp. 229–243). Springer Open.

Schoenfeld, A. H. (2014). What makes for powerful classrooms, and how can we support teachers in creating them? A story of research and practice, productively intertwined. *Educational Researcher*, *43*(8), 404–412.

Superfine, A. M. C. (2009). Planning for mathematics instruction: A model of experienced teachers' planning processes in the context of reform mathematics curriculum. *The Mathematics Educator*, *18*(2), 11–22.

Switzer, J. M. (2015). Which lake is bigger? *Teaching Children Mathematics*, *22*(4), 208–212.

van Es, E. A., & Sherin, M. G. (2008). Mathematics teachers' "learning to notice" in the context of a video club. *Teaching and Teacher Education*, *24*(2), 244–276.

Windschitl, M., Thompson, J., & Braaten, M. (2020). *Ambitious science teaching*. Harvard Education Press.

3

MENTOR TEACHERS' FOUR INTERTWINED ROLES WHEN LEADING LESSON STUDY WITH MATHEMATICS PRESERVICE TEACHERS

Gloriana González, Omar Hernández-Rodríguez, and Wanda Villafañe-Cepeda

Teacher educators are paying increased attention to leading lesson study with preservice teachers (PSTs) because the process of learning to teach is anchored in authentic classroom-based experiences. Lesson study supports teacher agency, and lesson study leaders establish routines for sustaining collaboration. Nevertheless, it can be challenging to promote agency and collaboration when leading lesson study with PSTs new to teaching and lesson study. We use data from a project where three host teachers in a field placement, namely, "mentor teachers," each led a lesson study team with preservice mathematics teachers enrolled in a university-based methods course. The mentor teachers had expertise in supervising field experiences but were new to lesson study. The project, which took place in Puerto Rico, aimed to bridge methods courses and field experiences through lesson study. We ask, "*How did the mentor teachers lead the teams?*" Additionally, we identify tensions between leading lesson study and supervising field experiences. Drawing from cultural-historical activity theory (CHAT) (Engeström, 1999), we create a model for leading lesson study with PSTs.

Our Innovation

In our innovation, each mentor teacher led a lesson study team with two to three PSTs. The PSTs were mostly in their second year of a 4-year teacher education program and did not have prior field experiences. Each team conducted two lesson study cycles. During the first cycle, each mentor teacher led their respective team to *prepare* for lesson study. Each team *planned* a research lesson to develop students' mathematical proficiency (Kilpatrick et al., 2001). The research lesson needed to include interactive technology introduced in the methods course. As part of the *study* and *plan* steps, the PSTs conducted three observations in the assigned mentor teachers' classrooms, focused on the mathematics proficiency framework.

DOI: 10.4324/9781003326434-5

The PSTs used evidence from the observations to foster students' mathematics proficiency when planning the research lesson. The mentor teachers *taught* the research lessons to their students. During the *reflect* step, each team revised the lesson. Then, a pair of PSTs *taught* the revised lesson to another class. Prior to the implementation of the innovation, the mentor teachers participated in one lesson study cycle to become familiar with lesson study. During the school year, the mentor teachers held a study group with the research team. In those meetings, we discussed a common agenda for implementing lesson study, offered suggestions for leading lesson study, examined the mathematics proficiency framework, and considered interactive technology software features. The study group discussions are beyond the scope of this chapter but reveal our intention to create a learning space for leading lesson study in relation to the aims of the methods class.

Four Roles When Mentor Teachers Lead Lesson Study

We identified four roles when mentor teachers led lesson study: Facilitator, knowledgeable other, team member, and supporter. These roles have been enacted by distinct individuals in other implementations of lesson study (Clivaz & Clerc-Georgy, 2020).[1] We found that the mentor teachers enacted these four roles simultaneously.

Mentor Teachers as Facilitators

According to Lewis (2016), there is a symbiotic relationship between facilitators and teachers during professional development with in-service teachers. Since lesson study is teacher-led, facilitation skills are emergent and co-constructed. Consequently, paying attention to the lesson study team's immediate needs is essential. For example, the facilitator needs to manage the time during the meetings, assert teachers' agency, and build trust. At the same time, the facilitator must keep the lesson study cycle moving along. As facilitators, the mentor teachers in our study established collaboration norms, introduced protocols, fostered discussion, and anticipated student thinking. The mentor teachers helped the team achieve the goals of each lesson study step and their facilitation actions were critical for sustaining lesson study.

Mentor Teachers as Knowledgeable Others

According to Takahashi (2014), the knowledgeable other's expertise provides a different perspective that enriches the team's discussions. Watanabe and Wang-Iverson (2005) provide a different view by stating that the knowledgeable other assumes the stance of a peer who fosters collegiality. Nevertheless, a common idea is that the knowledgeable other's input is not aimed at evaluating the team. In our study, the mentor teachers shared their knowledge of the curricular standards. The mentor teachers also prompted the PSTs to include opportunities for students to explain their reasoning, apply various solution strategies, and rely on multiple representations. Additionally, the mentor teachers showed how to incorporate into the research lesson the teaching moves for using technology that had been taught in the methods class. At times, the mentor teachers engaged the PSTs in short rehearsals that were like a mock-up lesson (Lewis et al., 2019). Overall, the mentor teachers used their knowledge to engage the team in considering pedagogical aspects relevant to the research lessons.

Mentor Teachers as Team Members

According to Clivaz and Clerc-Georgy (2020), when facilitators assume the group member's role, they show their vulnerability, for example, by stating that they do not know the solution to a mathematics problem. As team members, the mentor teachers had the responsibility of *teaching* the first implementation of the research lesson. In other steps, the mentor teachers assumed duties that could have been performed by any other team member, such as recording notes. Additionally, the mentor teachers positioned the PSTs as peers, fostered collaboration, and encouraged the team. In general, as team members, the mentor teachers modeled interactions within a professional learning community and communicated to the team that all contributions were valued.

Mentor Teachers as Supporters

The mentor teachers in our study had prior experience supervising PSTs. Nevertheless, leading lesson study required changes in the mentor teachers' typical patterns of interaction as supervisors. In our study, the mentor teachers identified teaching strengths and suggestions for improvement during the post-lesson *reflection* by giving feedback to the PSTs who taught the lesson. At the same time, the mentor teachers demonstrated empathy by acknowledging the PSTs' apprehensions about teaching and addressing their concerns. Moreover, the mentor teachers commended the PSTs' growth, a crucial action for building capacity to improve instruction (Lewis et al., 2009). Instead of evaluating the PSTs' capabilities, the mentor teachers supported them and situated the teams' goals in relation to mathematics education reform efforts in Puerto Rico. Additionally, the mentor teachers positioned the PSTs as agents of change by including new expectations for students to use technology and engage in mathematics discourse, thus reaffirming their decision to become mathematics teachers. In contrast with the knowledgeable other role which focused on pedagogical content knowledge, the supporter role entailed instilling enthusiasm for teaching and providing socio-emotional support.

Enacting the Roles during Lesson Study

We analyzed the data from four lesson study cycles in three lesson study teams, each led by a mentor teacher during the 2021 spring semester: Ms. López (7th-grade teacher), Mr. García (10th-grade teacher), and Mr. Martínez (8th-grade teacher).[2]

Each team implemented two lesson study cycles.[3] We focus on the steps where the mentor teachers led discussions, namely, *prepare*, *study, plan*, and *reflect*.[4] We relied mainly on the videos from the meetings. The research team segmented the videos into a timeline according to intervals (see Herbst et al., 2011). We coded the intervals for evidence of the mentor teachers' enactment of the four roles. For each role, we created profiles listing the actions within each lesson study step.

Prepare

As facilitators, the mentor teachers coordinated important logistics. Specifically, they shared the session agenda, elicited ideas for elaborating participation norms, distributed lesson study roles, modeled how to take minutes, established the means of communication, scheduled meetings, and set up the technology. The mentor teachers also fostered a collaborative environment by establishing expectations for participation, encouraging a positive attitude, and stating the importance of teamwork. The mentor teachers framed the lesson study activities in terms of larger professional goals, such as focusing on students who would ultimately benefit from the research lessons. The mentor teachers shared their learning about interactive mathematics technology and their experiences teaching mathematics. With these actions, the mentor teachers welcomed the PSTs to the teaching profession and created a sense of community.

At the same time, the mentor teachers' references to professional experiences positioned them as knowledgeable others. The mentor teachers fostered the use of technology in teaching, referred to mathematics standards, and situated the research lessons within the curricular sequence. The mentor teachers also provided descriptions of the classes where the research lessons would be implemented. The mentor teachers assumed the role of a team member by using inclusive language and committing to follow the participation norms established, setting the tone for future interactions. At the same time, the mentor teachers' backgrounds as field experience supervisors permeated during the *prepare* step. For example, the mentor provided suggestions to reduce the PSTs' concerns about teaching and assigned a written reflection about teamwork. With these actions, the mentor teachers positioned themselves as supporters.

Study

The mentor teachers' actions as facilitators focused the teams' examination of materials for the research lessons. The teams analyzed curricular resources (e.g., textbooks and standards), identified the mathematics topic for the research lesson, and established the technological features that would be available for their students. As knowledgeable others, the mentor teachers demonstrated uses of mathematics technology tools, navigated curricular materials, unpacked mathematical ideas, and provided examples of mathematics applications, thus relying on their *mathematical knowledge for teaching* (MKT; Ball et al., 2008) and *technological pedagogical content knowledge* (TPACK; Rosenberg & Koehler, 2015). The mentor teachers offered their rationale for their instructional decisions and acted as facilitators seeking the PSTs' justifications for their instructional decisions. There were limited instances where the mentor teachers assumed the role of a team member during the *study* step, such as when sharing an anecdote about a discovery lesson. The mentor teachers' role as supporters emerged when they held all accountable for studying the mathematics content of the lesson and when they were ready to step in to teach if one team member would be ill with COVID-19.

Plan

The mentor teachers assumed the facilitator and knowledgeable other roles in asserting the PSTs' agency while drawing on their own expertise. As facilitators, the mentor teachers continued to attend to logistical issues. At the same time, the mentor teachers fostered the development of a professional community and held the PSTs accountable for planning the research lesson. For example, the mentor teachers established goals and expectations, distributed pending tasks, identified who would teach the research lesson, reassured the team members' assistance to the assigned teacher, and elicited contributions. These actions created a collaborative environment. Another set of actions in the facilitator role involved the development of an inquiry stance. The mentor teachers prompted the PSTs to justify their decisions. Additionally, all the mentor teachers asked to see the slideshow from the students' point of view and engaged the team in "predictive planning" (Collet, 2019), a process by which a lesson study team anticipates students' work during the research lesson.

As knowledgeable others, the mentor teachers provided insights based on their MKT. For example, the mentor teachers situated the topic of the research lesson within the curricular sequence, emphasized students' development of conceptual understanding, broadened the multiple representations in the lesson, considered students' prior knowledge, and added scaffolds. The mentor teachers also identified relevant mathematics terms, explained mathematical ideas, and showed their appreciation for mathematics concepts. At the same time, the mentor teachers referred to the documents and ideas introduced in the methods class. Although we observed fewer actions as team members in relation to the other two roles, the mentor teachers continued to position themselves as one among the team. For example, everyone shared editing privileges when working on the lesson plan. Notably, the mentor teachers showed their vulnerability (e.g., dealing with technology problems). At the same time, the mentor teachers enacted their supporter role by approving the research lesson. Overall, the mentor teachers sought to create a research lesson that followed the specifications of their typical lessons while also complying with the methods' course expectations.

Reflect

The mentor teachers' actions were instrumental for holding post-lesson discussions. As facilitators, the mentor teachers started the *reflect* step by asking the

PSTs to offer their insights from teaching the lesson. The common agenda for the post-lesson discussions suggested to start with the insights from the PSTs teaching the lesson and then shift to the observers' comments. As facilitators, the mentor teachers also probed for the rationale for instructional decisions, elicited further comments, and linked the teams' observations to lesson revisions. Additionally, they assisted in preparations for implementing the research lesson again through rehearsals illustrating how the lesson would unfold. The mentor teachers also summarized and celebrated the teams' accomplishments. As knowledgeable others, the mentor teachers' actions were evenly split between the goals of discussing the implementation and revising the research lesson. The mentor teachers centered the teams' discussions of evidence of student thinking around mathematical proficiency. At the same time, the mentor teachers offered specific feedback to improve the research lesson by including multiple representations, adding scaffolds, changing the sequence of activities, and addressing student errors. The mentor teachers' feedback included using technology features to make student thinking visible and increase interactivity, thus drawing on their TPACK. The mentor teachers positioned themselves as team members by reflecting on ways in which they had incorporated technology into their teaching and sharing their difficulties. By providing reassuring comments, the mentor teachers aimed to boost the PSTs' confidence. As team members, the mentor teachers assessed the PSTs' wellbeing. In their supporter role, the mentor teachers provided feedback, including improvements to questioning techniques and strategies to maximize student engagement.

Examples of the Mentor Teachers' Roles when Leading Lesson Study[5]

We share instances illustrating how the mentor teachers enacted the roles of facilitator, knowledgeable other, and team member. The first instance is from the team led by Ms. López at the beginning of the second meeting in the *prepare* step. As stated in the common agenda shared, Ms. López discussed the Lesson Plan Template with the lesson study team.

> In the detailed part of the activities, it is important to distinguish between what is [a technical interruption] the launch, the exploration, and the summary. And in Part III that talks about the specific mathematics concepts that will be addressed, it is important that even though we will work on [solving] equations by addition and subtraction, it is important [...] there are some concepts that are going to be developed that are adding the opposite, the, the property—the inverse property of addition, the inverse property of multiplication, concepts like the reciprocal, the elements—the identity element for addition and as well as for multiplication. Those are important concepts that will be included in our lesson.
> [En la parte detallada de las actividades, es importante que se distinga entre lo que es [una interrupción técnica] la actividad de inicio, la actividad de

desarrollo y la actividad de cierre. Y en la Parte III que se habla de los conceptos matemáticos específicos que se trabajarán, es importante que, aunque nosotras vamos a trabajar con ecuaciones de suma y resta, […] ahí hay unos conceptos que se van a estar trabajando que son la suma del opuesto, la, la propiedad—la propiedad inversa de la suma, la propiedad inversa de la multiplicación, conceptos como el recíproco, los elementos—el elemento identidad tanto para la suma como para la multiplicación. Esos son conceptos importantes que van a estar incluidos en nuestra lección.]

Ms. López seamlessly shifted from the facilitator role to the knowledgeable other role. As a facilitator, Ms. López referred to the template, which helped the planning of the research lesson. As a knowledgeable other, Ms. López asked for specific descriptions of the different parts of the lesson. Additionally, Ms. López listed the target concepts and identified other relevant concepts, situating the research lesson within the mathematics curriculum.

The second instance illustrates the mentor teachers' simultaneous enactment of the roles of facilitator, knowledgeable other, and team member during planning. In their third meeting, Mr. García led a PST, Roberto, to *plan* the research lesson targeting the angle bisector theorem.[6] As a team member, Mr. García provided Roberto access to the textbook. As a knowledgeable other, Mr. García showed Roberto how to navigate the curricular materials. Additionally, Mr. García unpacked the theorem and showed connections with other theorems. Mr. García discussed another theorem about the concurrency of the medians of a triangle and discussed its application about a triangle's center of gravity. When talking about the concurrency of medians, Mr. García showed enthusiasm for mathematics (i.e., "Did you see that it looks very nice?"; "¿Viste que se ve bien bonito?"). Similarly, Mr. García led Roberto to construct a triangle and its orthocenter (i.e., the point of concurrency of the triangle's altitudes) using GeoGebra, a dynamic geometry software. Following Mr. García's instructions, Roberto dragged the vertices of the triangle and noticed that the orthocenter is a geometric invariant. Roberto reacted with excitement (i.e., "How nice!"; "¡Qué chévere!"), thus developing an appreciation of mathematics. During these interactions, Mr. García also enacted the facilitator's role by leading the planning process.

The third instance pertains to the *reflect* step in the team led by Mr. Martínez. Upon questions by a research assistant who at this moment assumed the facilitator role, the PST who taught the lesson asked Mr. Martínez for advice about how to handle a situation when students do not understand something despite many attempts to provide an explanation. Mr. Martínez validated the PST's concern and restated the objective of finding the recursive formula for a linear pattern. Then, Mr. Martínez explained the teaching dilemma.

Then, sometimes one does not find how to answer that, and the response is difficult because you don't want to move and jump from topic to topic either, that is, jump, tan, tan, tan, and attend to everything at the same time.

[Entonces a veces uno no encuentra cómo contestarle eso y se le dificulta la respuesta porque tú tampoco quieres moverte y brincar de temas, o sea, brincar, tan, tan, tan, y atender todo a la vez.]

Mr. Martínez provided various alternatives to manage the dilemma.

… well there you could then either incorporate a keyword or some key things that would give them a clue or at least tell them, "Look, okay, well do not worry. From this piece what I need you to understand is this, so that when we go to the next section you can understand that other piece."

[…pues ya ahí tú pudieses entonces o incorporar una palabrita clave o algunas cositas claves que les den una idea o al menos poder decirles, "Mira, okay, pues no se preocupen. De esta parte lo que necesito que entiendan es esto, para que cuando vayamos a la próxima sección entonces puedan entender esa otra parte."]

Finally, Mr. Martínez gave a suggestion.

In summary, the ideal is that you always know more content beyond what you are discussing. That is going to help you.

[Así que, en resumen, lo ideal es que siempre estés con contenido más allá del que estás discutiendo. Eso te va a ayudar.]

Mr. Martínez explained that teaching the recursive definition was for students to develop a conceptual understanding of linear patterns, which would help them derive the explicit formula. As a team member, Mr. Martínez demonstrated empathy and framed the situation as one typical in teaching ("sometimes one does not find how to answer that, and the response is difficult"). As a knowledgeable other, Mr. Martínez provided alternatives such as using keywords and focusing on one topic. The instances illustrate how the mentor teachers assumed various roles simultaneously, drawing on their knowledge and expertise.

Leading Lesson Study in Relation to the CHAT Framework

The CHAT framework provides a model of an activity system and is usually depicted with a set of triangles. The activity system involves a community, subjects who participate in the community, and an object (i.e., the aim of the activity). Wake et al. (2016) used the CHAT framework to analyze lesson study planning sessions, identifying the value of creating a lesson plan for teacher professional learning. Our object was to lead lesson study with PSTs. The community comprised the lesson study team. The subjects were mentor teachers. We identified as artifacts the physical or conceptual things that the mentor teachers used to lead lesson study. The rules included implicit and explicit ways of interacting. The division of labor refers to assignments that were spontaneously assumed or corresponded to lesson study roles. Figure 3.1 shows a model of the activity system for leading lesson study.

The artifacts, rules, and division of labor overlapped various roles. For example, the mentor teachers used curricular resources to position themselves as knowledgeable others drawing on their expertise in navigating curricular materials (Remillard, 2005). At the same time, curricular resources enabled the mentor teachers to behave as team members. Similarly, as team members, the mentor

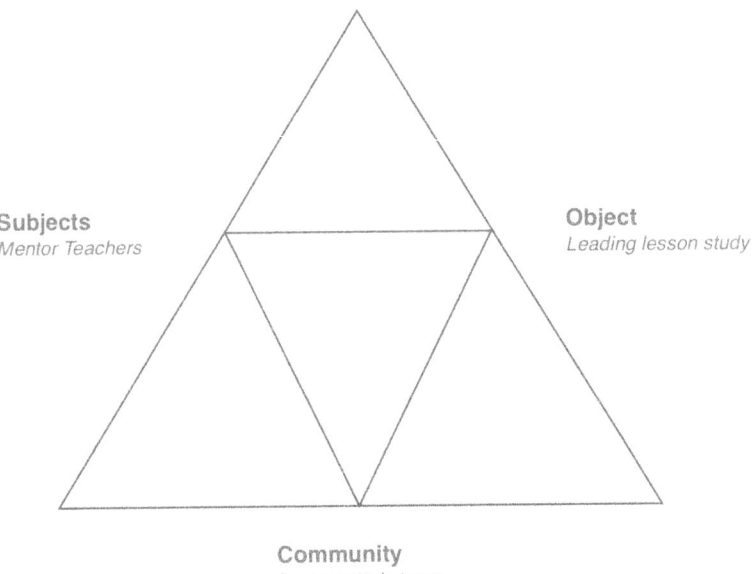

Artefacts
- Curricular resources (e.g., Teacher's edition of the textbook, standards)
- Methods course materials (e.g., Lesson plan template, rubric of moves for teaching with technology, strands of math proficiency)
- Online interactive platform

Subjects
Mentor Teachers

Object
Leading lesson study

Community
Lesson study team

Rules
- PSTs are responsible for lesson study involvement (e.g., draft the research lesson, edit slides, study math content, ready to teach)
- Mentor teachers are responsible for delivering high-quality instruction
- Mentor teachers support PSTs in complying with methods course requirements

Division of Labor
- PSTs undertake assigned lesson study roles
- Mentor teachers approve lesson study activities
- Mentor teachers accountable for creating a professional community

FIGURE 3.1 Activity System When Leading Lesson Study.

teachers modelled how to undertake some of the lesson study roles. At the same time, the mentor teachers were responsible for completing the lesson study steps and approving the research lessons. The mentor teachers referred to the methods course materials to plan the research lessons, thus bridging the methods course and the field experiences. Ultimately, in the division of labor, the mentor teachers were accountable for creating a professional community. An underlying rule of providing high-quality instruction solidified the lesson study team's sense of belonging to a professional community.

The mentor teachers' experiences as field experience supervisors sometimes posed challenges to leading lesson study. For example, typical discussions during the *reflect* step pertain to evidence of student thinking. However, the mentor teachers sometimes provided feedback to the PSTs about their performance. Another example is related to the division of labor. In some teams, the PSTs created a draft of the research lesson, while in other teams, the mentor teachers provided the draft. Further assistance may be needed for mentor teachers to overcome usual patterns of interaction while also drawing on their knowledge.

Concluding Thoughts

Our identification of the various roles that the mentor teachers enacted during lesson study illustrates the challenges of leading lesson study. The mentor teachers built on the methods course and drew on their knowledge as field experience supervisors to attend to the PSTs' needs. The mentor teachers demonstrated empathy, reassured the PSTs, and provided strategies to manage the uncertainty of teaching. Moreover, the mentor teachers created a reassuring environment. The mentor teachers empowered the PSTs by drawing from their knowledge as supervisors and instilled in them a sense of belonging to the teaching profession which was evident by their commitment to instructional improvement, sense of collegiality, and unwavering encouragement despite the pandemic-related challenges.

Future research should examine mentor teachers' roles when leading lesson study. This research could be the basis for professional development opportunities for mentor teachers. Acknowledging mentor teachers will ultimately result in improved support for PSTs. By drawing on mentor teachers' knowledge, researchers can help to bridge methods courses and field experiences toward more cohesive teacher education programs.

Acknowledgments

The work was supported by a project funded by the National Science Foundation, Division of Undergraduate Education (#1930950 & #1930971 granted to Omar Hernández-Rodríguez, PI; Wanda Villafañe-Cepeda, Co-PI; and Gloriana González, PI). Any opinions, findings, conclusions, or recommendations presented are only those of the investigators and do not necessarily reflect the views of the

National Science Foundation. Thanks to Yency Choque-Dextre and Juliette M. Moreno Concepción for their research assistance.

Notes

1 Researchers have documented other roles as well, such as convenor of lesson study and researcher (see Clivaz & Clerc-Georgy, 2020). Nevertheless, we focus on three afore-mentioned roles (facilitator, knowledgeable other, and team member), with distinct attention to their typical role of field experience supervisors, which is analogous to that of "teacher trainer" as specified by Clivaz and Clerc-Georgy (2020).
2 We use pseudonyms for study participants.
3 An exception was Mr. García who co-taught the first version of the research lesson with a PST.
4 We did not analyze the teaching step because the mentor teachers did not lead a discussion during this step.
5 We translated the excerpts that were originally in Spanish to English, considering the context of the discussion and colloquial terms in Puerto Rico. Transcription conventions include em dash, "—," for self-interruptions and "[...]" for text that we deleted for brevity.
6 Mr. García's team started with two PSTs. However, one of them withdrew from the course. The planning session included one PST.

References

Ball, D. L., Thames, M. H., & Phelps, G. (2008). Content knowledge for teaching: What makes it special? *Journal of Teacher Education, 59*(5), 389–407.

Clivaz, S., & Clerc-Georgy, A. (2020). Facilitators' roles in lesson study: From leading the group to doing with the group. In *Stepping up lesson study* (pp. 86–93). Routledge.

Collet, V. (2019). *Collaborative lesson study*. Teachers College Press.

Engeström, Y. (1999). Activity theory and individual and social transformation. In Y. Engeström, R. Miettinen, & R.-L. Punamäki (Eds.), *Perspectives on activity theory* (pp. 19–38). Cambridge University Press. https://doi.org/10.1017/CBO9780511812774.003

Herbst, P., Nachlieli, T., & Chazan, D. (2011). Studying the practical rationality of mathematics teaching: What goes into "installing" a theorem in geometry? *Cognition and Instruction, 29*(2), 218–255. https://doi.org/10.1080/07370008.2011.556833

Kilpatrick, J., Swafford, J., & Findell, B. (Eds.). (2001). *Adding it up: Helping children learn mathematics*. National Academy Press.

Lewis, J. M. (2016). Learning to lead, leading to learn: How facilitators learn to lead lesson study. *ZDM, 48*(4), 527–540. https://doi.org/10.1007/s11858-015-0753-9

Lewis, C., Friedkin, S., Emerson, K., Henn, L., & Goldsmith, L. (2019). How does lesson study work? Toward a theory of lesson study process and impact. In R. Huang, A. Takahashi, & J. P. da Ponte (Eds.), *Theory and practice of lesson study in mathematics* (pp. 13–37). Springer.

Lewis, C. C., Perry, R. R., & Hurd, J. (2009). Improving mathematics instruction through lesson study: A theoretical model and North American case. *Journal of Mathematics Teacher Education, 12*(4), 285–304.

Remillard, J. T. (2005). Examining key concepts in research on teachers' use of mathematics curricula. *Review of Educational Research, 75*(2), 211–246. https://doi.org/10.3102%2F00346543075002211

Rosenberg, J. M., & Koehler, M. J. (2015). Content and technological pedagogical content knowledge (TPACK): A systematic review. *Journal of Research in Technology Education*, *47*(3), 186–210. https://doi.org/10.1080/15391523.2015.1052663

Takahashi, A. (2014). The role of the knowledgeable other in lesson study: Examining the final comments of experienced lesson study practitioners. *Mathematics Teacher Education and Development*, *16*(1), 4–21.

Wake, G., Swan, M., & Foster, C. (2016). Professional learning through the collaborative design of problem-solving lessons. *Journal of Mathematics Teacher Education*, *19*(2–3), 243–260. https://doi.org/10.1007/s10857-015-9332-9

Watanabe, T., & Wang-Iverson, P. (2005). The role of knowledgeable others. In P. Wang-Iverson, & M. Yoshida (Eds.), *Building our understanding of lesson study* (pp. 85–91). Research for Better Schools.

Lesson Study Steps in Preservice Teacher Education

4

THE PREPARE STEP

Setting the Stage for Successful Lesson
Study with Preservice Teachers

Jenifer Hummer and Kristin Lesseig

When mathematics and science educators engage in lesson study, it is critical to allocate sufficient time and resources to the *prepare* step. As described by Lewis and colleagues (2019, p. 18), the goal of *prepare* is to "establish a lesson study team that is valued by its members and has reasonably efficient processes for learning together." *Prepare* is often overlooked by those planning or facilitating lesson study as aspects of the step overlap and are interrelated to the *study* step. In this chapter, we draw explicit attention to the *prepare* step with a focus on recommendations for preparing preservice teachers (PSTs) to participate in lesson study. We describe how others have chosen to (1) introduce PSTs to lesson study; (2) develop cooperative learning norms; (3) attend to the formation of research teams; and (4) negotiate with field partners the target standards and instructional materials the PSTs will use to design their instruction. For each of these aspects, we include our recommendations to best support the goals of *prepare* and maximize learning for PSTs.

The descriptions and recommendations throughout the chapter reflect our synthesis of research on conducting lesson study with PSTs who are preparing to teach mathematics or science at the K-12 level. We also analyzed documents submitted by 15 Lesson Study for Mathematics and Science Teacher Educators Conference (LSMSTEC) participants who have conducted lesson study with PSTs. Specifically, the LSMSTEC teacher educators shared how the schedule and agenda were formed, norms were established, and roles were assigned. We individually reviewed each document and recorded details for each category before working together to note overall themes.

DOI: 10.4324/9781003326434-7

Introducing Preservice Teachers to Lesson Study

One criticism of lesson study as practiced outside of Japan is its lack of grounding in lesson study principles and philosophy (Seleznyov, 2018; Yoshida, 2012). In Japan, lesson study is an integral part of teachers' professional learning, beginning in initial teacher training and continuing throughout their careers. Japanese teachers thus enter the process with a firm commitment to continuous improvement and view the research lesson as a place to test out their collective theories about teaching (Fujii, 2014).

In the United States, lesson study is still a relatively novel approach within teacher education programs. Thus, teacher educators must dedicate adequate time to introducing PSTs to the goals, process, and history of lesson study. Without this foundation, there is a risk of attending to only the superficial features of lesson study and/or modifying the process in ways that detract from the long-term learning goals. A critical mistake is to think the goal of lesson study is to develop an exemplary lesson rather than to support instructional improvement through teacher learning and a commitment to use one's classroom as a research site (Stigler & Hiebert, 2016; Yoshida, 2012). Such misunderstandings may lead to an overemphasis on lesson revision and reteaching, thus limiting time spent discussing the relationship between student thinking and the instructional context. Moreover, an emphasis on immediate responses or revisions may unintentionally derail efforts to develop productive dispositions toward learning from teaching, an important long-term goal of teacher education and lesson study.

Lesson study within initial teacher preparation is predominantly initiated by university instructors and embedded in a practicum experience or methods course. Instructors often introduce PSTs to lesson study through articles or other texts that describe the history and format of lesson study (e.g., Lewis, 2002; Stigler & Hiebert, 1999; Takahashi & Yoshida, 2004). PSTs' introduction to lesson study may also include opportunities to view and discuss videos of Japanese lessons or Japanese lesson study (e.g., Parks, 2008; Ricks, 2011). Ideally, instructors use these readings and videos to launch a discussion around the process, goals, and benefits of lesson study. Instructors may also model the lesson study process by conducting an abbreviated lesson study within the university course. This may include peer teaching or a demonstration lesson. These activities introduce PSTs to foundational principles of lesson study and support the development of productive norms for analyzing and discussing teaching.

Unlike in Japan, lesson study is not fully embedded in the professional learning culture of US teachers. Thus, everyone involved in the process, including university faculty and school partners, needs to learn about the goals and structures of lesson study. The importance of introducing lesson study principles not only to PSTs but also to mentor teachers was a clear theme that emerged from our analysis of LSMSTEC conference documents. A LSMSTEC participant described potential benefits of running a parallel lesson study with mentor teachers who could then be tapped to facilitate lesson study with PSTs. This seems like a promising model, though as the participant admitted, such an approach might be difficult to sustain without grant support to pay for substitute teachers and compensate lead teachers for time spent planning and coordinating future lesson studies for PSTs.

In the literature, we found two cases that discussed intentional preparation for participants outside of the university methods course. Bjuland and Helgevold (2018) held workshops with university faculty and school-based teacher mentors. Together, they developed a "handbook of lesson study" with suggestions for questions mentors might pose to PSTs, and other details related to observing teaching and making predictions during the planning of the research lesson. Amador and colleagues (Amador & Carter, 2018; Amador & Weiland, 2015) provided classroom teachers with opportunities to evaluate videos of lesson study debriefing conversations and discuss ways to focus on student reasoning during lesson analysis meetings. School partners were provided with a manual for field experiences including a three-step protocol for engaging in lesson study.

We recommend that those organizing and facilitating lesson study with PSTs also read guide books and research articles about lesson study and participate in lesson studies to deepen their understanding of the history and theoretical underpinnings of lesson study. When lesson study is new to the teacher educator charged with implementation, we suggest seeking outside support from someone experienced in lesson study, if possible, who can provide an initial orientation (Rasmussen, 2016). With a firm grounding in lesson study principles, teacher educators will be better equipped to avoid common pitfalls such as inadvertently focusing on *teachers* rather than *teaching* or viewing the lesson plan as a script to be followed. The importance of ensuring that *all* members of the lesson study community have a shared understanding of the goals and rationale behind each procedural element cannot be overstressed. We recommend that organizers plan well in advance for ways to incorporate mentor teachers, fellow faculty members, and others in this aspect of *prepare*.

Developing Cooperative Learning Norms

Despite the recognized importance of productive norms, we found few empirical studies that made explicit mention of either what or how norms for participating in lesson study were established. Perhaps because lesson study with PSTs is most often embedded in university courses, researchers assume general norms for equitable participation in class discussions are sufficient (e.g., Ricks, 2011). In the few studies in which norms were discussed, researchers highlighted the need for norms specific to lesson study. For example, in addition to establishing norms for collaborative discussions, researchers also attended to norms that might support PSTs in knowing what to notice in classrooms (Larssen et al., 2018).

LSMSTEC participants all stressed the importance of establishing norms, although some participants only noted that "discussion norms were established for the class." When specific norms were mentioned, they included professional norms (e.g., be curious, participate, and provide effective feedback) and norms specific to the lesson study environment. We consider this latter set of norms, such as norms to ensure that everyone contributes to the research theme and that the conversation focuses on student understanding, rather than evaluating the teacher or PST facilitating the lesson, as critical for successful lesson study.

In addition to listing what norms were important, LSMSTEC participants shed light on how norms were established. Several participants used specific protocols or tools to introduce lesson study norms such as those provided by the Lesson Study group at Mills College (lessonresearch.net). The group suggests the following process for collaboratively identifying norms that reflect the values of team members: (1) Individual reflection, (2) sharing and discussing potential norms, (3) synthesizing the agreed-upon norms, (4) recording those norms so they can be continually revisited, and (5) opportunities to practice these norms. Included tools provide suggestions for general norms to ensure a well-functioning group (e.g., share the air; listen thoughtfully with an open mind) as well as norms specific to the goals of lesson study (e.g., maintain a focus on student thinking). We recommend this as a useful starting point, especially for those new to lesson study.

We also recommend that norms be established early, revisited often, and reinforced through intentional actions by those facilitating the lesson study. Lesson study norms, and processes for establishing and maintaining them, should be linked to the principles and goals of lesson study. One LSMSTEC participant described how the course instructor and PSTs collaboratively set norms after reading about lesson study. In our own work, we first establish that the purpose of lesson study is to use a classroom lesson as a space to collectively learn about students' thinking and instructional practices that honor and advance their thinking. We then agree upon ground rules for how we share and respect others' ideas while simultaneously challenging those ideas and generating alternatives. To thwart teachers' and PSTs' tendency to evaluate the teacher facilitating the research lesson, we consistently reinforce the view that teaching is about decision-making. Our discussions then focus on the use of evidence to consider the affordances and constraints of the teaching decisions that unfold during a lesson. This shift helps move conversations away from evaluation toward a more productive inquiry stance that supports further learning (Cochran-Smith & Lytle, 1999).

Successful lesson study depends on productive cooperative learning norms. Spending time early in the cycle to establish group norms, coupled with continual monitoring and reinforcement of those norms, can support more efficient and effective ways of learning together. Through these efforts, teacher educators can avoid two challenges to successful lesson study: Managing participation and managing time (Lewis et al., 2019). The first challenge refers to making sure all team members feel included and that no voice habitually dominates the discussion. The second challenge addresses the need to keep conversations focused on critical aspects of teaching and learning in order to make the best use of the time provided and meet the long-term goals of lesson study.

Attending to the Formation of Research Teams

The formation of the research team is a critical element to a successful lesson study. Traditionally, lesson study research teams are formed through district-wide initiatives or through voluntary teams of teachers who want to work together to meet an instructional goal (e.g., Lewis & Perry, 2014). These teacher teams are supported by a facilitator and knowledgeable others (e.g., an instructional coach or outside mentor) who provide content expertise and/or feedback on pedagogy (Baldry & Foster, 2019; Lewis et al., 2019). In contrast, PST lesson study teams are typically formed through course structures with methods course instructors planning and leading the lesson study teams (e.g., Akerson et al., 2017; Erbilgin & Arikan, 2021). Field placements concurrent with teaching methods courses may influence the formation of lesson study teams as PSTs at the same placement may be grouped together (e.g., Amador & Carter, 2018; Chassels & Melville, 2009). Our analysis of the LSMSTEC documents revealed similar trends, with PST lesson study teams heavily influenced by methods courses and field placement assignments. As such, the lesson study team sizes varied from three to 15 PSTs, with most teams comprising three to six PSTs.

While the formation of lesson study teams for PSTs is somewhat simplified and structured based on methods courses and field placements, we recommend that mathematics and science teacher educators carefully consider how teams are formed. The strengths of individual PSTs should be taken into account, as the PSTs will serve as resources to each other (e.g., Lewis & Perry, 2014; Suh & Seshaiyer, 2015). For example, PSTs with advanced subject matter knowledge can be grouped with PSTs who need more content support. This grouping may open opportunities for all participants to deepen their understanding of the mathematics or science principles embedded in the lesson if members are held accountable to providing explanations and ensuring all members can access the ideas. The same argument holds for pedagogical knowledge. Grouping PSTs with varying pedagogical expertise and experience has the potential to increase the pedagogical knowledge of the team members as participants may attend to different facets of instruction.

In addition to attending to participants' differing levels of content and pedagogical content knowledge, power structures within lesson study should be considered (Baldry & Foster, 2019). Given PSTs are students and novice teachers, the university instructor or mentor teacher could easily become the dominant voice in the lesson study group. A goal of lesson study is that teachers drive the learning (Lewis et al., 2019) with all members seen as valuable contributors to the process. In teacher preparation, PSTs, mentor teachers, and university instructors should have equal opportunities to engage as experts and learners. More specifically, PSTs should drive the learning while being supported by facilitators and knowledgeable others. Carefully attending to the cooperative learning norms and the dynamics of a research group will support the success of a lesson study.

To support efficient lesson study processes, Lewis and colleagues (2019) recommend establishing group roles at the beginning of the study. Group roles can support engagement and more equitable participation among lesson study members. The Lesson Study Group at Mills College (2022) suggests the following roles: Facilitator, notetaker, recorder, timekeeper/process checker, and a logistics coordinator. These roles can be determined by the methods instructor or other lesson study team members. Within the LSMSTEC data, ten out of 15 groups followed the group roles suggested by Lewis and colleagues (2019). Of these ten groups, five mentioned that group roles rotated and/or that the PSTs decided on all roles except the facilitator. The methods course instructors facilitated the lesson study for ten out of 15 teams. For the remaining teams, one group was facilitated by the mentor teacher, another by a PST after the initial lesson was planned, and in three cases, the role of facilitator rotated among group members. While there are various approaches to assigning and maintaining group roles across the literature and LSMSTEC data, we recommend that the group roles rotate. Rotating roles may mitigate the negative influence of power structures and ensure that all team members experience leadership opportunities. Doing so is especially important for the facilitator as that person may be perceived to have the most power.

If an outside knowledgeable other is included in the lesson study, their role can be critical to the success of lesson study. The knowledgeable other can support learning by providing content and/or pedagogical expertise. PSTs and mentor teachers may be intimidated by a knowledgeable other, so participants may hesitate to share their thinking during lesson study. Because of this power dynamic, Baldry and Foster (2019) suggested that the university instructor serve as the knowledgeable other and lead facilitator so long as that person is adept with the lesson study process. We also suggest that mentor teachers may play the role of a knowledgeable other if, in addition to content and/or pedagogical expertise, they have experience with lesson study and are supported by the university researchers (e.g., Amador & Carter, 2018). Before mentor teachers serve as knowledgeable others, we recommend they are oriented to lesson study as described above and participate in lesson studies that are modeled by university researchers or outside lesson study experts.

Negotiating with Field Partners

Partnerships between the university and field placement schools should also be carefully negotiated. This is especially important given that lesson study is not as well known outside of Japan. In the US, for example, the school structure and culture do not always support the time and space necessary for lesson study. Additionally, PSTs often have busy schedules with their coursework and other obligations (Chassels & Melville, 2009). Hence, working with partner schools and mentor teachers for scheduling and developing learning norms is critical. The power relationships within these structures should also be considered. For example, often, the lesson study topics or goals are imposed upon the mentor teacher and the PSTs by university instructors (e.g., Akerson et al., 2017; Baldry & Foster, 2019). To mitigate some of these challenges, mentor teachers should be brought into the curricular negotiations early to help identify standards and select instructional materials. Additionally, to build sustainable relationships with partner schools, methods instructors can work with mentor teachers and school leaders to decide on an instructional goal or topic to guide the lesson study process (e.g., Bjuland & Helgevold, 2018; Erbilgin & Arikan, 2021; Ricks, 2011). Some lesson study teams have allowed PSTs to choose the research question for the lesson study (e.g., Bjuland & Helgevold, 2018; Parks, 2008). However, because there are often curricular demands (e.g., district pacing guides) for mentor teachers, we recommend that methods instructors work closely with mentor teachers to negotiate the goals of a research lesson. Not only does this create a more inclusive working relationship, but the mentor teacher and school leadership will be more likely to buy into lesson study when the work supports the goals of the classroom and school (Takahashi & McDougal, 2016).

Along with negotiating the lesson study goals and research questions, school structures and cultures must be considered (Takahashi & McDougal, 2016). Schools outside of Japan often do not have dedicated professional development time or structures in place to allow substitute teachers to cover classes. In contrast, Japanese teachers are expected to spend time in lesson study meetings or observations as part of their professional work. Based on these differences, Takahashi and McDougal (2016) identify a number of elements necessary for successful lesson study in the US. These include a school principal who supports lesson study and communicates lesson study goals to faculty; at least one other school leader (e.g., teacher, assistant principal) who is an advocate for lesson study; a school-wide goal for lesson study that is compelling for teaching and learning and; a school administration committed to providing resources for sustainable lesson study (e.g., funding and time). To ensure these elements exist, school partnerships should be established early so that school leaders have a voice in all aspects of the *prepare* step.

One way to strengthen these partnerships and support schools is to position PSTs and universities as resources. Schools can look to universities to provide instructional resources and expertise in terms of the latest research on effective mathematics and science teaching and learning. Lesson study with PSTs and mentor teachers is a type of professional development that universities can provide with no financial obligations to the school. When the *prepare* step and other lesson study steps are executed effectively, this low-cost professional development can continue as PSTs and mentor teachers transition into leadership roles and can lead lesson study once the university methods instructor is no longer available. These ideas align well with the recommendations from Takahashi and McDougal (2016) who noted that lesson study is more likely to continue at the school when there's a strong advocate for lesson study such as a school leader or teacher.

Concluding Thoughts

Through our review of the literature, the LSMSTEC data, and reflection on our own experiences with teaching methods courses, we noted several challenges to conducting lesson study with PSTs that need to be addressed during *prepare*. For instance, the non-voluntary nature of the experience (i.e., when embedded into student teaching, methods courses, seminars, or other program courses/requirements) brings possible challenges due to differing motivations and time constraints. Finding time to conduct a full lesson study cycle within a single semester or methods course may be difficult. There are also power dynamics that need to be considered. These power issues may surface at two levels. At the lesson study group level, there is a danger of mentor teachers or university facilitators being positioned as authority figures. Second, there is potential conflict between the goals of the university and K-12 school mandates. Rather than imposing their own goals upon schools, university lesson study facilitators need to consider how to leverage the schools as equal partners to create a mutually beneficial learning environment.

While finding time for lesson study and addressing power relationships within a methods course is challenging, the structure helps expedite some of the critical aspects of *prepare*. For one, collaborative learning norms are often embedded into the university course norms. Through other method's course activities, the PSTs will also likely be accustomed to discussing teaching practice, so goals of lesson study related to continuous improvement and learning from teaching will be reinforced. Overall, we see lesson study as an opportunity to support PSTs beyond the methods course by helping them move beyond the goal of developing "the perfect" lesson to becoming practitioners and scholars of continuous improvement.

Here we summarize our recommendations for *prepare* to maximize the learning potential of conducting lesson study with PSTs:

1 Introduce PSTs to the purpose of lesson study by using readings, videos, and outside support (e.g., lesson study experts). Design parallel activities to ensure school partners and other participants have a shared understanding of lesson study goals and procedures.
2 Attend to cooperative learning norms—including professional norms to ensure equitable participation and norms specific to observing teaching. Even when norms have been established among PSTs through coursework, these norms need to be reintroduced and reinforced when beginning to work with mentor teachers and other school partners.
3 Consider power dynamics and group roles to ensure the learning norms can be sustained and learning opportunities are optimized. All lesson study participants should have chances to share ideas and contribute to the group learning.
4 Navigate school culture to ensure collegial partnerships. University faculty should collaborate with school leaders and mentor teachers to develop lesson study research goals and instructional materials.

The *prepare* step is critical to establishing a foundation for efficient and effective lesson study. Differentiating *prepare* highlights and names the necessary "hidden labor" that teacher educators do to make lesson study with PSTs successful. Unique circumstances should be considered for PSTs as learning goals of lesson study may differ from those for in-service teachers. In particular, while a lesson study goal of learning from practice should be emphasized across all experience levels, PSTs may need focused support in learning from practice. Laying a foundation in learning from practice is likely to develop PSTs who continue to learn from their teaching and engage in continuous improvement as lifelong learners in their fields. To achieve this, university researchers and instructors need to consider school partners as true collaborators to establish productive norms and a shared sense of the process, philosophy, and goals of lesson study.

References

Akerson, V. L., Pongsanon, K., Park Rogers, M. A., Carter, I., & Galindo, E. (2017). Exploring the use of lesson study to develop elementary preservice teachers' pedagogical content knowledge for teaching nature of science. *International Journal of Science and Mathematics Education, 15*(2), 293–312.

Amador, J. M., & Carter, I. S. (2018). Audible conversational affordances and constraints of verbalizing professional noticing during prospective teacher lesson study. *Journal of Mathematics Teacher Education, 21*(1), 5–34.

Amador, J., & Weiland, I. (2015). What preservice teachers and knowledgeable others professionally notice during lesson study. *The Teacher Educator, 50*(2), 109–126.

Baldry, F., & Foster, C. (2019). Lesson study in mathematics initial teacher education in England. In R. Huang, A. Takahashi, & J. P. da Ponte (Eds.), *Theory and practice of lesson study in mathematics* (pp. 577–594). Springer.

Bjuland, R., & Helgevold, N. (2018). Dialogic processes that enable student teachers' learning about pupil learning in mentoring conversations in a lesson study field practice. *Teaching and Teacher Education, 70*, 246–254.

Chassels, C., & Melville, W. (2009). Collaborative, reflective and iterative Japanese lesson study in an initial teacher education program: Benefits and challenges. *Canadian Journal of Education/Revue Canadienne De l'éducation, 32*(4), 734–763.

Cochran-Smith, M., & Lytle, S. L. (1999). Relationships of knowledge and practice: Teacher learning in communities. In A. Iran-Nejar, & P. D. Pearson (Eds.), *Review of research in education* (pp. 249–305). AERA.

Erbilgin, E., & Arikan, S. (2021). Lesson study to support preservice elementary teachers learning to teach mathematics. *Mathematics Teacher Education and Development, 23*(1), 113–134.

Fujii, T. (2014). Implementing Japanese lesson study in foreign countries: Misconceptions revealed. *Mathematics Teacher Education and Development, 16*(1), 65–83.

Larssen, D. L. S., Cajkler, W., Mosvold, R., Bjuland, R., Helgevold, N., Fauskanger, J., & Norton, J. (2018). A literature review of lesson study in initial teacher education: Perspectives about learning and observation. *International Journal for Lesson and Learning Studies, 7*(1), 8–22.

Lewis, C. (2002). Lesson study: A handbook of teacher-led instructional improvement. Research for Better Schools.

Lewis, C., Friedkin, S., Emerson, K., Henn, L., & Goldsmith, L. (2019). How does lesson study work? Toward a theory of lesson study process and impact. In R. Huang, A. Takahashi, & J. P. da Ponte (Eds.), *Theory and practice of lesson study in mathematics* (pp. 13–37). Springer.

Lewis, C., & Perry, R. (2014). Lesson study with mathematical resources: A sustainable model for locally-led teacher professional learning. *Mathematics Teacher Education and Development, 16*(1), 22–42.

Parks, A. N. (2008). Messy learning: PSTs' lesson-study conversations about mathematics and students. *Teaching and Teacher Education, 24*(5), 1200–1216.

Rasmussen, K. (2016). Lesson study in prospective mathematics teacher education: Didactic and paradidactic technology in the post-lesson reflection. *Journal of Mathematics Teacher Education, 19*(4), 301–324.

Ricks, T. E. (2011). Process reflection during Japanese lesson study experiences by prospective secondary mathematics teachers. *Journal of Mathematics Teacher Education, 14*(4), 251–267.

Seleznyov (2018). Lesson study: An exploration of its translation beyond Japan. *International Journal for Lesson and Learning Studies, 7*(3), 217–229.

Stigler, J., & Hiebert, J. (1999). *The teaching gap: Best ideas from the world's teachers for improving education in the classroom.* The Free Press.

Stigler, J. W., & Hiebert, J. (2016). Lesson study, improvement, and the importing of cultural routines. *Mathematics Education, 48*(4), 581–587.

Suh, J., & Seshaiyer, P. (2015). Examining teachers' understanding of the mathematical learning progression through vertical articulation during lesson study. *Journal of Mathematics Teacher Education, 18*(3), 207–229. 10.1007/s10857-014-9282-7.

Takahashi, A., & McDougal, T. (2016). Collaborative lesson research: Maximizing the impact of lesson study. *ZDM, 48*(4), 513–526.

Takahashi, A., & Yoshida, M. (2004). Ideas for establishing lesson-study communities. *Teaching Children Mathematics, 10*(9), 436–443.

The Lesson Study Group at Mills College. (2022). *Prepare Your Team.* https://lessonresearch. net/conduct-a-cycle/prepare-your-team//

Yoshida, M. (2012). Mathematics lesson study in the United States: Current status and ideas for conducting high quality and effective lesson study. *International Journal for Lesson and Learning Studies, 1*(2), 140–152.

5

THE STUDY STEP

Building Preservice Teachers' Knowledge in Content and Pedagogy

Rachelle Meyer Rogers, Ryann N. Shelton, and Trena L. Wilkerson

According to Lewis et al. (2019), the *study* step has two main goals: (1) Establish a lesson study team where all members are valued and uphold processes for learning together and (2) establish a focus for the lesson study and to build team members' knowledge about the topic (Lewis et al., 2019). While Lewis et al. (2019) pointed to the *study* step as being a singular phase in the lesson study process, participants of the Lesson Study for Mathematics and Science Teacher Educators Conference (LSMSTEC) identified distinct sub-phases that should be used with preservice teachers (PSTs) in both the *prepare* step (described in the previous chapter) and the *study* step, which is the focus of this chapter. The *study* step, as identified by the LSMSTEC participants, consists of (1) upholding established norms for learning together, (2) establishing a research focus, (3) selecting a content topic, and (4) building content knowledge about the topic.

When examining how LSMSTEC participants, who conducted lesson studies in various K–12 settings with PSTs attending different universities, engaged in the *study* step, there appeared to be a variety of approaches. For some, the *study* step consisted of identifying a research theme and selecting a content topic; however, no time was spent on studying the academic content and teaching materials. For others, the *study* step involved identifying a content topic and engaging in a study of the topic, but not a research theme. Another alternative for the *study* step consisted of selecting a content topic without taking time to study the content or identify a research theme. They also focused on various components such as content, standards, research, existing lessons, frameworks, resources, tasks, materials, curriculum, and teaching and learning practices when supporting PSTs in the *study* step. In our examination of the various approaches to the *study* step, the variety of these approaches was partly attributed to the different settings in which the PSTs were engaging in lesson study, including in different types of courses or focused

DOI: 10.4324/9781003326434-8

on different aspects related to teaching and learning. In this chapter, we share our perspectives of the *study* step after participating in LMSTEC, an overview of our revised *study* step with PSTs conducting lesson study in our teacher education program, and five recommendations related to conducting the *study* step with PSTs.

Our Examination of the *Study* Step

After LSMSTEC participants shared small-group presentations about lesson study at their universities, conference attendees gathered in a whole group to reflect on the *study* step and on engaging PSTs in this step. In identifying the similarities and differences between our experiences with the *study* step and Lewis et al.'s (2019) strategies, we, as three middle and secondary mathematics teacher educators at a private university in central Texas, left with additional questions about the *study* step. First, we wondered if the LSMSTEC-defined *study* step is consistent with the Lewis et al. (2019) study step. Second, we wondered what resources should be utilized with PSTs during the *study* step. Third, we wondered who makes up the research team when lesson study is conducted with PSTs, specifically considering the role of the university instructor and mentor teacher. Last, and perhaps most significant, we wondered if the *study* step is essential in lesson study regardless of who participates. Throughout this chapter, we address our wonderings and provide recommendations for implementing the *study* step when engaging PSTs in lesson study.

Setting the Stage for the *Study* Step

It is critical that the cooperative learning norms established in the *prepare* step are reinforced throughout the *study* step. The social dynamics and the manner of talking and responding to others' comments are crucial in becoming a recognized team member and a critical element to engaging in the *study* step (Lewis et al., 2019). As highlighted in the previous chapter, establishing equal power relationships among lesson study members is important (Lenski et al., 2018; Sjunnesson, 2020) for all members to be seen as valuable contributors. If individuals feel their values, experiences, and interests are invited and encouraged to be heard within the group, they are more invested in engaging in and contributing to the *study* step. Because PSTs are novice teachers, they can view themselves as less powerful than university instructors or mentors, which according to Lenski et al. (2018) can cause an imbalance of contributions in discussions. Those who view themselves as less powerful may remain silent or become complacent. Therefore, to avoid a position of power or authority in the lesson study team, we recommend university instructors or mentor teachers support PSTs by serving as a knowledgeable other during the *study* step.

The Context

Our middle and secondary mathematics PSTs engage in lesson study in their senior year internship, which consists of a 1-year clinical experience at either a middle or a secondary professional development school. PSTs engaged in the 1-year clinical experience were assigned to a mentor teacher and took on the role of co-teacher throughout the academic year. As part of the clinical experience, PSTs attend weekly seminar classes on the university campus where they were introduced to lesson study, planned for the experience, and implemented their research lesson in a professional development school. After the LSMSTEC, we considered how we were conducting lesson study with our middle and secondary mathematics PSTs, particularly in the *study* step. We realized that we did not spend enough time focused on the background knowledge of content, misconceptions, and content connections. Therefore, we refocused our *study step* in light of our new knowledge and wonderings when conducting lesson study with two groups of PSTs: One middle grades group and one secondary group. We offer our reflections and recommendations in the following sections.

Engaging PSTs in *Study*

Establishing a Research Theme

Based on our examination of the literature, LSMSTEC discussions, and our own experiences with lesson study, we recommend the *study* step begin with university instructors reminding PSTs of the established group norms and introducing the task of establishing a research theme. While there is no widely shared definition of a research theme, Takahashi and McDougal (2016) defined it as a broad teaching-learning goal that moves beyond a particular topic, a desired outcome for students. To establish a research theme, members of the lesson study team identify outcomes they would like their students to have, what they have been able to achieve, and how they can help close the gap (Lewis, 2002; Takahashi & McDougal, 2016). Identifying a research theme is an important element because it serves as a motivator that should influence the third step of the lesson study cycle, the planning of the lesson. For example, if teachers want their students to become problem-solvers and communicate their thinking, then the instructional plan should consist of a problem-based learning approach encouraging exploration and communication among students instead of a direct teaching approach (Meyer & Wilkerson, 2011).

In order for PSTs to develop a research theme, we utilized a specific tool provided by the Lesson Study Group at Mills College (2013). We asked PSTs to consider three questions while thinking about the students they teach during their internship:

1 What qualities would you like your students to have 5 years from now or when they graduate from your institution?
2 What qualities do your students have now?
3 Are there any gaps between the qualities you want your students to have and the qualities they currently possess? What are the gaps you would most like to address?

We then asked PSTs to compare the ideal qualities (revealed in responses to question 1) and the actual student qualities (revealed in responses to question 2) and share the information with their lesson study team members. Team members looked for commonalities to identify a group research theme before determining a topic focus for their research lesson. The research theme that emerged for the middle grades group was: *To become independent thinkers and resilient problem solvers that communicate their mathematical understanding with confidence.* The research theme that emerged for the secondary group was: *For students to develop resilience in problem-solving through positive collaboration and to improve communication skills to excel in society.* Both the middle and secondary PST groups had developed research themes revolving around developing students who were independent thinkers, resilient problem-solvers, and able to communicate confidently. It is important to note that we did not participate in establishing the research theme or the planning of the lesson to avoid introducing a position of power or authority in the lesson study team.

Establishing a Content Focus

Once all lesson study members agree upon the research theme, time should be spent identifying and investigating the topic of study. It is important that PSTs are again reminded of the established group norms before engaging in this process to reinforce equitable participation. Teams often begin by reviewing the content scope and sequence to see when various topics are scheduled to be taught at their assigned grade levels. They might even review students' assessment performance and identify a topic based on needs. We, however, found that the two factors that have the biggest influence on PSTs identifying a content focus are the scope and sequence for when a topic is scheduled to be taught and/or the grade level or subject matter. For example, the middle grades PSTs decided to focus on a sixth-grade mathematics lesson on mean, median, mode, and range because they all had prior experience teaching that topic and it was scheduled to be taught at the time the research lesson was to be implemented. The secondary-level PSTs decided to teach a lesson in algebra to work with students in a grade level that some PSTs had not experienced during their preparation programs.

Building PSTs' Knowledge

Once the topic is identified and agreed upon by all team members, the next task is for the team to strengthen their knowledge of the topic by engaging in a careful study of content and teaching materials (Lewis et al., 2019; Takahashi & McDougal, 2016). According to the National Research Council (2001), teachers must be knowledgeable in their content and knowledgeable in their content pedagogy. Lesson study supports the investigation of a content topic allowing participants to gain a deeper knowledge of the topic (Guner & Akyuz, 2020) and a deeper knowledge of teaching in the content area (Meyer & Wilkerson, 2011).

To best support the study of academic content and teaching materials, lesson study participants should value the knowledge and experiences of all team members. To this end, all PSTs contributed by sharing resources and actively listening to their team members' experiences with the identified topic and being open to all instructional ideas. PSTs, however, are novices and may not be aware of high-quality content and teaching resources that could be utilized to build their knowledge. As previously mentioned, to avoid a position of power or authority in the lesson study team, we recommend the university instructor should serve as the knowledgeable other during this part of the *study* step. The university instructor has extensive knowledge of the content and teaching materials and can help support the learning process. Having the university instructor serve in this role helped us answer our wondering as to if a knowledgeable other was needed and when they might be introduced. Likewise, we determined that the mentor teacher should also serve as a knowledgeable other after the research lesson has been implemented and team members are engaged in discussions on the lesson. The mentor teacher works at the school and is most familiar with the students and the curriculum.

During the LSMSTEC, participants reported a variety of ways to support PSTs as they developed their knowledge base around the content focus: Existing lessons, frameworks, research findings, tasks, materials, curriculum, and teaching and learning practices. It was clear to us that no two teacher educators used all the identified components or used the components in the same way, which lead to our initial wondering about what resources the knowledgeable other should provide to PSTs during the *study* step. Consequently, we intentionally introduced a variety of research-based resources and instructed PSTs to concentrate on five areas as identified in Figure 5.1, to assist in building their content and pedagogical knowledge. While there are numerous high-quality content materials that enable a team to move beyond what members already know, we utilized several resources from the National Council of Teachers of Mathematics and other research-based resources to support PSTs in developing their selected concept, in relation to the five areas. The purpose was to provide high-quality, classroom-based resources that addressed effective mathematics teaching practices to engage all students and provided background in deep content understanding. The PSTs were instructed to explore the provided resources along with others to which they have access, to gather information and develop an understanding of both the content and mathematical practices related to their content focus. We provide a framework of the five elements to help the PSTs organize their thinking (Figure 5.1).

As PSTs developed their understanding of the content focus, they were encouraged to examine their state standards starting with the grade level that addressed their content focus, and then review both prior knowledge (area 1) and future connections (area 2) to see what considerations were needed to inform their instructional decision-making. We noted that PSTs tend to focus on a single lesson or skill prematurely in the process. We recommend that the knowledgeable other support

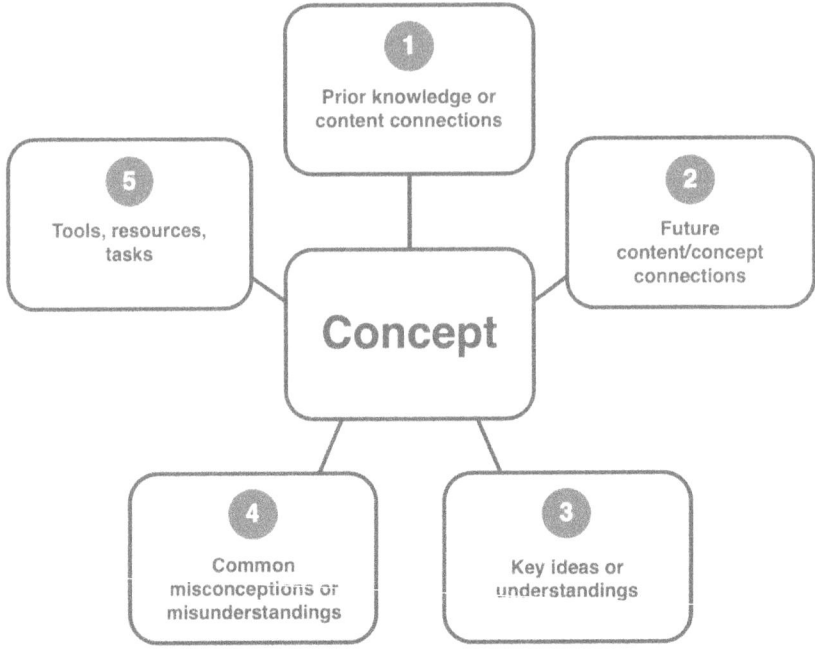

FIGURE 5.1 Five Areas of Focus on Content.

the process by encouraging team members to focus on the larger unit of study and keep the research theme in mind.

During this process, the PSTs crafted questions and areas to consider as they examined the materials and processed ideas. For example, the team focused on developing independent thinkers and discussed if they wanted the mathematics to be done in a group or individually, and if done in a group, how could they ensure that each student was thinking on their own? PSTs perused journal articles, books, and other research-based resources to identify key ideas or understandings (area 3) and common student misconceptions and misunderstandings (area 4) associated with the selected content focus. Finally, they explored a variety of instructional tools, tasks, and resources along with mathematical tasks (area 5) that would support their students in understanding the topic and help the students develop strong mathematical practices.

As part of the content focus process, Lewis et al. (2019) recommended lesson study team members identify something they would like to personally learn to support deeper content or pedagogical knowledge. In examining the experiences of the LSMSTEC participants, having PSTs set personal learning goals was not common. We were curious as to how PSTs would engage in setting a personal learning goal and implemented this strategy with our PSTs. We discovered that our PSTs identified personal learning goals connected to the five areas of focus on content

(Figure 5.1), which resulted in more reflection or gained insights from the investigations. An example of a personal learning goal from one PST was to engage in a vertical alignment review. Another PST identified her personal learning goal by stating it was insightful to consider students' prior knowledge and what they would need to know for the following year.

Based on the examination of the literature, LSMSTEC discussions, and our personal experiences, we recommend PSTs be asked to identify personal learning goals because it highlights an integral part of the *study* step, to build knowledge. Setting a personal learning goal can serve as a motivator for PSTs to have an authentic interest in the work. We recommend PSTs set their personal learning goals prior to engaging in the examination of academic content and teaching materials. While the framework's focus on content provides a general guide for material exploration, PSTs' personal goals provide individual incentives to parallel the general material exploration. PSTs could be encouraged to identify personal learning goals connected to content or pedagogical knowledge, and the university instructor or knowledgeable other may need to give examples of learning goals that reflect an inquiry related to the team's work (Lewis et al., 2019).

Recommendations

We left the LSMSTEC with several wonderings. We wondered about what resources should be provided for PSTs during the *study* step, about the role of the university instructor and the mentor teacher, and whether all the goals of the *study* step were essential. Through our review of the literature, the LSMSTEC data, and examination of our own experiences with conducting lesson studies with PSTs, we believe that in order for PSTs to build their knowledge in content and pedagogy, it is essential to implement the goals of the *study* step. The *study* step consists of: (1) Upholding established norms for learning together, (2) establishing a research focus, (3) selecting a content topic, and (4) building content knowledge about the topic. When PSTs do not experience the *study* step in the lesson study cycle, PSTs may not experience lesson study in the way it is intended. The *study* step provides PSTs with the opportunities to have explicit discussions about content and pedagogy and to focus on students' learning.

In reflection, we share five recommendations related to conducting the *study* step with PSTs. First, we believe it is essential for teacher educators to ensure lesson study team members uphold the collaborative expectations for the learning community established in the *prepare* step throughout the *study* step. For example, all lesson study participants have valuable experiences to share, and all experiences should be heard and considered (Lenski et al., 2018). Second, it is important to allow PSTs to develop a research theme that describes a desired outcome for the students they work with during their clinical experiences. Allowing PSTs to establish the research theme gives them voice and ownership (Lenski et al., 2018) and serves as an important motivator in closing the gap between where their students are and where they want them to be

(Takahashi & McDougal, 2016). Third, PSTs should identify the content topic while keeping the desired outcome for students in mind and identify a personal learning goal connected to content or pedagogical knowledge. Fourth, university instructors, who have extensive knowledge in content and content pedagogy, should serve as the knowledgeable other as PSTs engage in an intentional study of content and teaching resources. The knowledgeable other can provide quality content resources for PSTs to review while also allowing PSTs to identify specific resource(s) they identify as valuable in strengthening their own content and pedagogical knowledge. PSTs must feel they are contributing to the learning community and not feel overpowered by the knowledgeable other (Lenski et al., 2018). Fifth and finally, mentor teachers should serve as a second knowledgeable other later in the lesson study cycle when engaged in the post-lesson discussion. The mentor teacher best knows the participating students and the school curriculum.

Concluding Thoughts

The *study* step is an important component in the lesson study cycle. In previous years, we expedited this step to move PSTs into the planning step more quickly. However, our engagement in the LSMSTEC and our subsequent study highlighted the importance of spending time more fully developing the elements of the *study* step for our PSTs. Time is often a challenge, but we recommend that time should be taken to implement the goals of the *study* step.

With enough time to select a research theme, identify a content focus, and examine content and teaching materials, the *study* step provides the foundation and motivation to build team members' knowledge of the topic. When PSTs have strong content and pedagogical knowledge, they are better prepared to plan effective lessons (National Research Council, 2001), which directly supports the next step of the lesson study cycle, the *plan* step.

References

Murata A. (2011). Introduction: Conceptual overview of lesson study. *Lesson Study Research and Practice in Mathematics Education: Learning Together*. https://doi.org/ 10.1007/978-90-481-9941-9_1

Guner, P., & Akyuz, D. (2020). Noticing student mathematical thinking within the context of lesson study. *Journal of Teacher Education, 71*(5), 568–583. http://doi. org/10.1177/0022487119892964

Lenski, S. J., Rigelman, N. R., Bright, A. L., Thieman, G., & Ferner, B. R. (2018). What teacher educators learned about negotiating power relationships during lesson study planning. *Northwest Journal of Teacher Education, 13*(1), 1–15. https://doi.org/10.15760/ nwjte.2018.13.1.2

Lewis, C. C. (2002). *Lesson study: A handbook of teacher-led instructional change*. Research for Better Schools, Inc.

Lewis, C. C., Friedkin, S., Emerson, K., Henn, L., & Goldsmith, L. (2019). How does lesson study work? Toward a theory of lesson study process and impact. In R. Huang, A.

Takahashi, & J. Pedro da Ponte (Eds.), *Theory and practice of lesson study in mathematics: An international perspective* (pp. 13–36). Springer International Publishing AG.

Meyer, R., & Wilkerson, T. (2011). Lesson study: The impact on teachers' knowledge for teaching mathematics. In L. C. Hart, A. S. Alston, & A. Murata (Eds.), *Lesson study research and practice in mathematics education: Learning together* (pp. 15–26). Spring Science.

National Research Council & Mathematics Learning Study Committee. (2001). *Adding it up: Helping children learn mathematics.* J. Kilpatrick, J. Swafford, & B. Findell (Eds.). National Academy Press. https://doi.org/10.17226/9822

Sjunnesson, H. (2020). Initializing phase of lesson study: Communication a special didactic tool in mathematics. *International Journal for Lesson & Learning Studies, 9*(3), 261–275. https://doi.org/10.1108/IJLLS-02-2020-0007

Takahashi, A., & McDougal (2016). Collaborative lesson research: Maximizing the impact of lesson study. *Mathematics Education, 48*(4), 513–526.

The Lesson Study Group at Mills College. (2013). Can you lift 100 kilograms? The lesson research cycle. Retrieved from https://lessonresearch.net/content-resource/can-you-lift-100-kilograms/

6

THE PLAN STEP

Extending Learning to Preservice Teachers in a Full Lesson Study Cycle

Kevin J. Reins and Matthew Melville

In this chapter, we describe the essential features of the *plan* step in accordance with focus group exercises at LSMSTEC, reviewed literature, and our experiences to illuminate how teacher educators (TEs) can facilitate a successful *plan* step. We provide an in-depth look at enactments of lesson study among mathematics preservice teachers (PSTs) to illustrate how a methods course repositioned PSTs as main participants alongside in-service teachers rather than "predominantly in-service teachers with a few preservice teachers mixed in" (Lewis, 2019, p. 487). There are few examples in the existing literature of mathematics PSTs conducting a full lesson study cycle including a research lesson in K-12 schools (Larssen et al., 2018). To add to this literature, this chapter focuses on the *plan* step, while arguing that lesson study is a viable alternative to the common lesson planning instruction within methods courses. The emphasis on the plan step can allow TEs to better place faith in PSTs' abilities to anticipate student thinking or higher-level reflection.

The Plan Step

There are two main goals of the *plan* step: (1) Identify the learning goals for the research lesson and larger unit, based on standards, research, and students' current understanding, and (2) develop a lesson plan that bridges from students' current understanding to the identified new learning and think through or try out key portions (Lewis et al., 2019).

Many attempts to implement lesson study worldwide have occurred, with outcomes often falling short of expectations from observing Japanese lesson study. Often, failures occur by not implementing or poorly implementing the entirety of

DOI: 10.4324/9781003326434-9

lesson study, especially the *plan* step (Doig et al., 2011; Fujii, 2014). Although lesson plan creation is a beneficial artifact of the *plan* step, the purpose of lesson study is improving instruction more broadly through solving a teaching/learning problem (Watanabe et al., 2008).

During the *plan* step, teachers take the materials (*kyouzai¹*) and combine them with their understanding from *study* to make a high-quality lesson plan. Although this seems straightforward, like many aspects of lesson study, there are underlying features propelling the lesson plan into a tool that can improve teachers' mathematical instructional practices through highlighting ways to address a problem of practice. During the *plan* step, PSTs articulate broad, far-reaching goals aligned with a research theme. These goals are situated in a unit design among other types of goals to increase both content and pedagogical content knowledge.

After creating goals, the team then develops a detailed lesson plan. The intended audience for the plan is the observers and knowledgeable others who need to understand the group's reasoning behind their decision-making process (Melville, 2017). The plan includes a snapshot of the unit to situate where the students' current knowledge will be when the lesson occurs, its intellectual challenge (*kadai*), the task (*mondai*), anticipated student solution methods and possible misconceptions, how to tie the lesson together (*matome*), and what types of questions will be asked to prompt further inquiry (*hatsumon*). These descriptors of the sections of the lesson allow for teachers to have a common language to build off of and focus on a more specific problem of practice more than the lesson's logistical aspects. For example, teachers can specifically say they want to improve on the *hatsumon* to help students understand and engage in the task. This common language and the partitioning of the lesson enables a deeper level of conversation to occur.

To help provide more detail to their lesson plans and demonstrate every aspect has been fully considered, teachers provide snapshots of their boardwork plan (*bansho,* the art of organizing and displaying math solutions for discussion), hint cards or questions to elicit student thinking, and any other materials used. These detailed lesson plans are commonly viewed as the benefit of lesson study but are actually a representation of the teachers' completed work and development. They are a product of detailed instructional study and conversations held with other teachers to refine their ideas and thoughts.

Engaging PSTs in the *Plan* Step

In this section of the chapter, we examine the characteristics of the *plan* step with PSTs, explain its difficulties, and suggest ideas to other methods instructors for implementation. To identify these essential characteristics, we reviewed literature, artifacts, documented experiences, and group discussions from LSMSTEC.

Where PSTs Succeed and Fall Short in the United States

Complementing some conference learnings, literature supports the struggles that PSTs have in lesson study and how lesson study benefits their learning. For example, PSTs classroom inexperience may lead to difficulties when anticipating student responses and misconceptions about mathematical ideas (Burroughs & Luebeck, 2010). Yet, with support from TEs, PSTs changed their thinking about only classroom logistics, such as lesson pacing, to understanding student thinking when engaged in lesson study (Meiliasari, 2013). Additionally, PSTs develop their ability to work collaboratively with peers and other in-service teachers (Mostofo, 2013; Win, 2022).

There are three main areas during the *plan* step where PSTs need to develop their understanding: (1) Creation of high-quality tasks; (2) anticipating student thinking and misconceptions; and (3) having a complete understanding of the trajectory of the material (Burroughs & Luebeck, 2010; Fujii, 2014; Lewis et al., 2019).

PSTs and in-service teachers need to improve their capacity to anticipate student outcomes from tasks and make connections between the research lesson and previous student knowledge. These can be developed through time and learning from others; however, explicitly addressing these weaknesses is difficult (Burroughs & Luebeck, 2010). This may be the case because when planning, (1) teachers may fail to grasp students' current knowledge and design the lesson based on that presumption, (2) teachers may not anticipate student thinking in enough depth, accuracy, and breadth to write a plan likely to promote learning, and (3) teachers may focus on logistical elements rather than key student experiences producing tension or contradiction (Lewis et al., 2019).

Teachers in Japan refer to their national curriculum standards when preparing a lesson. (Melville, 2017). They view each lesson as a piece of the whole, and the entire unit plan is considered when planning lesson goals and desired outcomes (Doig et al., 2011). However, US teachers may fall short of this deep understanding if they begin lesson planning without first considering the unit design, the long-term content trajectory, and the lesson's role within the larger unit (Lewis et al., 2019). If teachers fail to identify the new learning that will occur during the lesson, their planning focuses on what they will do instead of what students will learn from the activity. One possible explanation for this discrepancy between Japanese teachers and US teachers is that in the US, teachers do not have a national curriculum.

Knowing about these difficulties, we suggest some ideas and provide a case demonstrating assisting PSTs fuller engagement in the *plan* step in the next section.

Learnings from Conference Collective Thought Exercises

From our analysis of conference artifacts, TEs engage PSTs in the *plan* step by urging them to consider the larger unit and ways to bridge the gap between students' current understanding and the lesson aim. This helps PSTs improve their ability to anticipate student thinking, possible student responses, possible student misconceptions, and make tasks more engaging for students. TEs use a variety of strategies, including rehearsal lessons among peers, providing input on documents/research lesson plans, and utilizing frameworks and thought-focusing templates. Through these processes, PSTs better document student learning and thinking by focusing on what students are learning rather than what they should be "doing" in class.

Despite these benefits, challenges remain. For instance, PSTs may need to change their own prior ideas about planning, and redirect their focus to student thinking, possible misconceptions, and how to activate students' prior knowledge when creating lesson plans. When TEs provide support to help alleviate these difficulties, PSTs can improve their planning practices.

A Case of Extending the Plan Step to PSTs in a University Methods Course

In this section, we expound the *plan* step by discussing practices and structures, including modifications to meet the needs of undergraduate PSTs as main participants in a full lesson study cycle. The setting and model maintain as much fidelity as possible in the all steps of the lesson study cycle, *prepare, study,* and *plan; teach;* and *reflect* (see Reins, 2020 for more details).

We will utilize Lewis and colleagues' (2019) two goals for the *plan* step to organize this section. The first goal, *Identify the learning goals for the research lesson and larger unit, based on standards, research, and students' current understanding,* provides focus for PSTs and will be summarized as (1) Setting Research Purpose and Goal-setting. The second goal, *Develop a lesson plan that bridges from students' current understanding to the identified new learning and think through or try out key portions,* outlines planning action and is subsequently summarized as (2) Creating Bridging Experiences.

(1) Setting Research Purpose and Goal-setting

PSTs need to understand the purpose and importance of goal setting for the lesson to deeply learn from the *teach* and post-lesson *reflect* steps. Therefore, it is necessary to spend time thoroughly discussing the students' learning experience. This discussion begins by defining the lesson's *clear research purpose,* that is, *What is the actual problem we are trying to solve?* This can stem from a teaching and learning problem, one that is mathematical in nature, or simultaneously both. Considering this should constitute discussions of two research objectives, (1) a potential Problem of Instruction (PoI), and (2) a shared, broad goal that goes beyond the PoI, topic, and grade level (a research theme).

At Kevin's university, 2 months are allocated in the secondary mathematics methods course, which typically enrolls between five and 13 undergraduates, for a full lesson study cycle on two PoIs. For the first research objective, the instructional problems are identified by in-service teachers. Rather than asking PSTs to produce these PoIs, Kevin discusses with the in-service educators possible topics for lesson study at the school year's beginning. Kevin guides the in-service educators to identify lessons that are the most difficult for students to understand, cause confusion, contain misconceptions, contain naïve ideas associated with content, or relate to topics from the Common Core State Standards for Mathematics that seem difficult to incorporate into lessons.

As these PoIs are shared with the PSTs, Kevin also explains that sometimes in lesson studies conducted in schools, teachers select favorite lesson ideas, lessons impressionable to observers, or conduct multiple iterations of a lesson to perfect it. Takahashi and McDougal (2016) state that these types of thinking ultimately miss out on lesson study's ideal purpose, conducting research to seek a solution to a current teaching/learning problem.

With this in mind, the in-service teacher's PoIs are narrowed and two are chosen by the PSTs, two lesson study teams of PSTs are formed, each with an in-service teacher (sometimes the same in-service teacher works with both teams), and questioning ensues. This may need to be modified for larger methods courses such that the lesson study groups remain in the 3–6 student range.

Two vitally important questions from Takahashi and McDougal (2016) set the stage for helping identify the research theme in the PoI and are located at the forefront of our LS Goal-setting Template in Outlining a Clear Research Purpose (Figure 6.1). This goal-setting *plan* template is introduced before the lesson study planning template described below.

Next, Kevin and the students discuss what research themes are desired for semester focus. Together, they identify a broad teaching and learning goal for student development that goes beyond the PoI, yet fits the school, class, or topic. These might include habits of mind, expert ways of thinking, productive mathematical dispositions, or focus on a Standard for Mathematical Practice from the Common Core State Standards Initiative.

For example, focusing on improving students' ability to construct a viable argument and critique the reasoning and arguments of others might be selected as a research theme. By defining the research theme, two more questions become important, *What is our desired outcome for students?* and *What is an entry point for achieving that outcome?*

As an entry point and desired outcome for constructing and critiquing arguments, a teacher might utilize large Think Board X2s adhered on acrylic boards (https://www.think-board.com/products/think-board-x2) to allow students to develop and share their mathematical work and representations. The *study* and *plan* steps might focus on learning how one selects which pathways/routes/solution strategies to discuss during the lesson and how their selection might be based on students' level of mathematical **reasoning**:

empirical: Evidence that supports but does not justify a conjecture,
preformal: The intuitive explanations and partial arguments that lend insight into
   what is occurring, or

LS Goal-setting Template

Problem of Instruction

What is your team's chosen Problem of Instruction (PoI)?

Outlining a Clear Research Purpose

Why do students typically struggle with XXX concept or skill?

How can we design a lesson so that students learn XXX concept or skill better than they have in the past?

What is our research theme?

   a) What is our desired outcome for students?

   b) What is an entry point for achieving that outcome?

Existing Goals & Standards for the Lesson

What are the goals and standards expressed in their text? By their teacher?

What are the CCSSM leading up to this topic, at grade level, and where is it going?

Coherence Maps may be useful: achievethecore.org/coherence-map

Related prior learning standards	Learning standards for this lesson	Related later learning standards

What are some initial draft ideas of your goals for the lesson? Write 2-3 for each.

What most teachers focus upon ...

   Content goals (CCSSM related and other mathematical content goals)

     •

     •

   Mathematical Practice Standard goals:
   Mathematical Practice Standards and Their Verbs

     •

     •

What Lesson Study wants us to emphasize ...

   Broad overarching content goals for the chapter:

     • The student understands that ...

     • The student will know that ... appreciate that ...

     • The student will be able to ...

   Affective goals (growth in interest, attitudes, emotions, feelings, behaviors):

     • e.g., Loving a challenge; internal motivation; excitement to learn about the lesson

     • e.g., Students will persevere and develop grit in solving challenging problems

   Connection goals:

   Mathematical Connections

     •

     •

   Real-world Connections

     •

     •

   Broad, long-term goals for student development:
   (habits of mind, expert ways of thinking, productive mathematical dispositions)

     •

     •

   Technology goals:

     • Use of manipulatives

     • Exploration with Applets/Online Tools

FIGURE 6.1 Lesson Study Goal-setting Template.

formal: Argumentation based on logic in determining mathematical certainty, making statistical inferences),

and ***sense-making***:

their understanding of a situation, context, or concept by connecting it with existing knowledge or previous experience.

During this aspect of the *plan* step, Kevin helps the PSTs plan for how they will assist their students in questioning their own and others' reasoning and sense-making by viewing one another's work. All student work is important as they ask others to explain explorations, conjectures at various levels, false starts, and partial explanations en route to a solution. It is these student-to-student interactions that provide opportunities for K-12 students to learn how to reason, sense-make, and comprehend the PST's desires for the lesson's mathematics.

Alternatively, a research theme could involve a new way of approaching a topic, something that will captivate students in a non-routine task (*mondai*), and intellectually challenge (*kadai*) students through a different instructional approach. Early in the semester, PSTs learn about theoretical approaches to teaching and learning mathematics (e.g., Realistic Mathematics Education and Embodied Mathematical Cognition through experiencing lessons and reading about each approach. One of these approaches may be an entry point for achieving an effective research theme. It is through the research theme that the effects of a chosen new approach would be studied.

Next, PSTs analyze existing goals for the lesson. They view the ready-made teacher textbook supplementals gathered during the *study* step and examine the existing goals suggested by the text and in-service teacher and how they mesh with the other lesson and chapter goals of the text's unit. Then, they examine the grade level Common Core State Standards for Mathematics leading to the content in the PoI, and their related later learning standards.

Teams utilize coherence maps of the Common Core (Achieve the Core, n.d.) https://achievethecore.org/coherence-map/ to find connections among the standards and existing knowledge. Once they understand the PoI, in-service teachers on each team are asked for any pre-existing student work so PSTs can analyze it for common errors, trends, and indicators of mis- or weak understandings, and placed in a Google Drive folder. These practices lessen weaknesses in the *plan* step identified by researchers.

Additionally, two to three students from grades 7–12 (as appropriate to the PoI) are recruited from an alternative section to the one used for conducting the lesson study for clinical interviews on the topic. The objective of these initial clinical interviews is to help PSTs grasp students' current knowledge and thinking about the PoI and utilize this data to write lesson goals. At this point, the questions inquire about how students think, rather than test lesson ideas. Testing comes later during rehearsals and in a second round of clinical interviews. The interviews help identify qualities that high school (HS) and middle school (MS) students are lacking that may become part of the research theme. Since researchers (Burroughs & Luebeck, 2010; Lewis et al., 2019) have noted that PSTs are weak at grasping students' existing knowledge and anticipating how they think, clinical interviews provide an opportunity for them to reflect on this as a team and bring these ideas into the *plan* step.

PSTs also utilize the *study* step findings to influence the lesson goals. Reins (2020) includes a modified template for what PSTs might be researching here;

examples include: *What are the in service teacher's past approaches and experiences with the topic and their successes and failures? What does current research on the teaching and learning of the topic provide?* If PSTs missed something crucial during their *study* step that might influence goal writing, Kevin assigns a research article on a finding, theory, or method. These readings generate new ideas that influence PSTs' thinking and may influence the lesson goals.

Within the LS Goal-setting Template, PSTs distill this information into a refined set of three to five lesson goals. These goals should help students understand something new, have a new ability, or cause them to feel differently about something, thus producing a cognitive change, i.e., a modification in the means and processes through which a person understands, acquires knowledge, or builds abstract structures from experience (Ibarra et al., 2020). The two lesson study teams share their work with the in-service teacher who suggested their PoI and get feedback. A knowledgeable other, such as Mathematical Sciences faculty members who have been introduced to lesson study and the role of the knowledgeable other, might also provide feedback.

(2) Creating Bridging Experiences

Each time the PSTs meet during the regularly scheduled methods course to plan, they complete an entry in the Planning Record Template to which the in-service teacher has access and can provide feedback. Here, they log all progress and work that was completed between and during class meetings. It also includes space for the next steps and goals at the end of the planning session.

As class periods are 50 minutes and time may be used to provide feedback or introduce items relevant to the PoI or research theme, the Planning Record Template helps PSTs generate a sense of accomplishment and determination to remain focused and continue moving forward. Each planning session begins with an assessment of the last meeting, what was to be accomplished between classes, and status. Five minutes at the end of each planning period are reserved for setting new goals for the next session and task assignments. These are written on a Planning Record Template:

- **Planning Date**/Class Date
- **Progress Made.** What was to be accomplished between meetings
- **Work Accomplished.** During the current meeting
- **Next Steps.** What and whom
- **Goals** for Next Planning Meeting

We, as Lewis and colleagues (2019), feel it is vital to utilize a lesson study planning template. Two widely used templates which work well with PSTs are The Lesson Study Alliance (LSA, no date) template (https://www.lsalliance.org/resources/) or the Mills College Teaching-Learning Plan Template (https://lessonresearch.net/

study-step/access-tlp/). These are periodically updated based on lived experiences and ongoing research, as evidenced by their revision history. Since lesson study is not strictly a linear process and there is no single correct way of navigating its sequentially outlined steps, the LSA group has a recommended flow for completion within the template. We loosely follow it as a guide but have also rearranged the order of attack when elements need revisiting.

The central point for PSTs to consider in planning their lesson is they are not focusing on finding materials and activities for student engagement, but rather how students experience the lesson's mathematics. All design choices for materials and engagement must lead toward developing and measuring the goals and research theme set forth for the lesson.

PSTs consider different stages of the lesson and how they might utilize *interaction problems* throughout it. An *interaction problem* is something that invites every participant to engage in meaningful mathematics learning; students cannot hide or be ignored. It develops agency, ownership, and identity as students explain and critique their ideas and recognize how they contribute to everyone's learning. An interaction problem, or a concatenated sequence of interaction problems, brings the mathematical idea, concept, or skill into the purview of the learner, to achieve the desired learning goals for cognitive change. For example, seeing which tilings of Algebra Tiles, and corresponding factored and general forms, produce results that are squares rather than rectangles, and determining how one might use these square-like equations to solve a quadratic equation that is not factorable is an interaction problem (Reins, 2014).

PSTs need solid examples of lessons that create tensions, contradictions, and drama within a lesson that will lead students, working as a team, to a breakthrough in their collective thinking. This is difficult for many PSTs, as some have not experienced this kind of instruction. The lessons modeled at the beginning of the semester have meaningful mathematics developed through sequences of interaction problems, so these provide experiences with several lessons developed in this way. This affords the opportunity to reflect and draw upon shared experiences within the methods course itself in attempting their own lesson's design in teams.

All lesson stages/elements must be considered individually and how they fit together to make a cohesive whole (*matome*), especially how its problems and segments encourage student reflection. In order for PSTs to anticipate student thinking and reflection with enough depth, accuracy, and breadth to write a plan likely to promote learning (Lewis et al., 2019), they conduct a second round of clinical interviews with two or three MS/HS students. Here, they test some of their ideas for sequencing and the interaction problems (*kodai*) they designed for the lesson, and interviews focus on how the intellectual challenge (*mondai*) of the interaction problems will reveal sequencing difficulties or problems in producing the line of thinking, sense-making, or activation of prior knowledge predicted.

Mock-up lessons or "rehearsals" can also be utilized to test lesson ideas. Rehearsals have one or two PSTs act as the instructors and teach (or co-teach) the lesson to a subset of their team (or to the other lesson study team) in the methods class, who assume the role of student participants. This is why two lesson study teams and PoIs are recommended, as PSTs can test their lesson ideas on each other in a novel way and experience what the design of their interaction problems and layout feels like from the students' viewpoint.

Their colleagues can explicate what they perceive and experience to help the other team make lesson improvements, with these rehearsals allowing PSTs to assess well-thought-out prompts, interaction problems, sequencing, their production of student engagement and thinking, instructional moves, *bansho* (organization of boardwork), and their data collection (actual results and whether quality evidence of learning goals are provided).

A final approach that aids PSTs in thinking through their lesson choices is asking a knowledgeable other, perhaps an accomplished K-12 educator or a university faculty member, to analyze the lesson's plan and offer suggestions. This is usually the same teacher or faculty member invited to the actual lesson study when taught by the PSTs in the participating in-service teacher's classroom. For more information on this collaboration, see Reins (2020).

How the Plan Step Influences the Remainder of the LS Process

The *plan* step as outlined prepares PSTs to focus deeply on how students experience mathematics and how their lesson design impacts all students' mathematical learning. This step sets the stage for the observers to ascertain what and how to provide feedback during the *teach* and the *reflect* steps of lesson study. This is allowed by giving the PSTs a full lesson plan and its one-page summary to the in-service teacher, invited knowledgeable others, and the other PST lesson study team prior to the lesson's teaching in the in-service teacher's classroom. Sometimes, this same lesson is taught with slight, real-time modifications to other class periods within the same day.

Concluding Thoughts

We recommend that other TEs who enact lesson study review our and other's essential features before engaging PSTs in the *plan* step. PTSs need to be viewed as capable leaders and provided a pathway for their enculturation into the discipline of good teaching through an in-depth engagement in the full practice of lesson study. Additionally, this careful planning prepares PSTs for success in all steps of the cycle and impacts their future planning of mathematics lessons.

Through engaging PSTs in the *plan* step of lesson study, this process will become natural to them, thus increasing their ability to continuously improve their instruction. The essential features of lesson study have often been described as an iceberg, with many of its features not immediately apparent; however, exposing them allows other cultures outside of Japan to better implement lesson study to its full extent (Hart et al., 2011).

Note

1 Words that are italicized parenthetically signify a Japanese term that loses meaning when translated. Therefore, we have kept their original language to describe aspects of lesson study.

References

Achieve the Core. (2023, June 22). *Coherence Map*. https://achievethecore.org/coherence-map/

Burroughs, E., & Luebeck, J. (2010). Pre-service teachers in mathematics lesson study. *The Montana Mathematics Enthusiast*, *7*(2&3), 391–400.

Doig, B., Groves, S., & Fujii, T. (2011). The critical role of task development in lesson study. In *Lesson study research and practice in mathematics education* (pp. 181–199). Springer.

Fujii, T. (2014). Implementing Japanese lesson study in foreign countries: Misconceptions revealed. *Mathematics Teacher Education and Development*, *16*(1), 65–83.

Hart, L. C., Alston, A. S., & Murata, A. (2011). *Lesson study research and practice in mathematics education*. Springer.

Ibarra, L., Soriano, A., Ponce, P., & Molina, A. (2020). The wit-learning methodology as a means for research skills acquisition: A longitudinal assessment. In M. Habib (Ed.), *Revolutionizing education in the age of AI and machine learning* (pp. 196–222). IGI Global. https://doi.org/10.4018/978-1-5225-7793-5.ch010

Larssen, D. L. S., Cajkler, W., Mosvold, R., Bjuland, R., Helgevold, N., Fauskanger, J., Wood, P., Baldry, F., Jakobsen, A., Bugge, H. E., Næsheim-Bjørkvik, G., & Norton, J. (2018). A literature review of lesson study in initial teacher education: Perspectives about learning and observations. *International Journal for Lesson and Learning Studies*, *7*(1), 8–22.

Lesson Study Alliance. (no date). Lesson Study Resources: Tools and useful documents: Template for a lesson research proposal. Retrieved on July 27, 2022 from http://www.lsalliance.org/resources/

Lewis, J. M. (2019). Lesson study for preservice teachers. In Huang R., Takahashi A., daPonte J. (Eds.), *Theory and practice of lesson study in mathematics: An international perspective (advances in mathematics education)* (pp. 485–506). Springer. https://doi.org/10.1007/978-3-030-04031-4_24

Lewis, C., Friedkin, S., Emerson, K., Henn, L., & Goldsmith, L. (2019). How does lesson study work? Toward a theory of lesson study process and impact. In R. Huang, A. Takahashi, J. daPonte (Eds.), *Theory and practice of lesson study in mathematics*. (pp. 13–37). Springer. https://doi.org/10.1007/978-3-030-04031-4_2

Meiliasari (2013). Lesson Study with Pre-service Teachers: Investigating the Learning of Pre-Service Teachers in Lesson Study Model of Teaching Practice Course. Fifth International Conference on Science and Mathematics Education. Penang, Malaysia.

Melville, M. D. (2017). *Kyozaikenkyu: An in-depth look into Japanese educators' daily planning practices*. Brigham Young University.

Mostofo, J. (2013). *Using lesson study with preservice secondary mathematics teachers: Effects on instruction, planning, and efficacy to teach mathematics*. Arizona State University.

Reins, K. J. (2020). Designing effective lesson study practices for mathematics education students. *Educational Designer: Journal of the International Society for Design and Development in Education*, *4*(13), 1–21.

Reins, K. J. (2014). Using tasks to assess and support student modeling, use of mathematical structure, and argumentation. *NEMJ: New England Math Journal*, *46*, 50–63.

Takahashi, A., & McDougal, T. (2016). Collaborative lesson research: Maximizing the impact of lesson study. *ZDM*, *48*(4), 1–14. 10.1007/s11858-015-0752-x

Watanabe, T., Takahashi, A., & Yoshida, M. (2008). Kyozaikenkyu: A critical step for conducting effective lesson study and beyond. *Inquiry into Mathematics Teacher Education*, *5*, 131–142.

Win, Y. M. (2022). Teacher educators' understanding of integrating lesson study into preservice teacher education. *Journal of Adult Learning, Knowledge and Innovation*, *4*(2), 52–61.

7

THE TEACH STEP

Engaging Preservice Teachers in Effective Practice

Rosemarie Michaels and Nicole Glen

The *teach* step of a lesson study cycle is the actual implementation and observation of a research lesson created to meet the academic and social needs of the students. In the *teach* step, one or more people teach while other adults observe and record students' learning (The Lesson Study Group at Mills College, 2022). Although the lesson is a carefully planned event attuned to student learning needs, the teacher implementing the plan must exercise professional judgment about what will be effective for the students, even if that means they stray from the lesson plan (The Lesson Study Group at Mills College, 2022). This chapter will overview the essential elements of the teach step and then outline two different ways that a teacher educator (TE) could accomplish this step with preservice teacher (PSTs).

Essential Elements of the *Teach* Step

According to Lewis et al. (2019), the goals of the teach step are to build habits of observation, test the team's hypothesis about learning, and enhance the quality and impact of educators' conversations by grounding them in a shared classroom experience. With PSTs, testing hypotheses about learning means facilitating the development of PSTs' pedagogical knowledge and pedagogy in mathematics and science and deliberately connecting coursework theory to actual teaching practice in K-12 classrooms. The collaboration between universities and schools is an important aspect of the teach step with PSTs. The experience of teaching or observing a research lesson with K-12 students provides authentic learning experiences and professional development opportunities for all participants (Burroughs & Luebeck, 2010; Cajkler et al., 2013; Michaels, 2015, 2020A, 2020B). In addition, the shared lesson study experience provides TEs and PSTs with a common reference that can be used to illuminate pedagogical theory during future class meetings. Therefore,

DOI: 10.4324/9781003326434-10

TEs should structure lesson study experiences during coursework experiences in partnership with K-12 classroom teachers.

Building Habits of Observation

Lewis and Hurd (2011) point out that lesson study is meant to focus on student learning and development and allow observers to be unencumbered by classroom management issues or real-time instructional decisions. This is different from typical observations of teachers in the United States, which tend to focus on a teacher's critique. The research lesson must be structured to allow for "as many windows as possible into student thinking" (Lewis & Hurd, 2011, p. 52). This can be accomplished with lesson activities where students express their thinking and justify their thoughts orally or in writing rather than rote memorization or worksheets. The data collected during the lesson's teaching supports a rich post-lesson discussion. It can be helpful to aid PSTs with forms, specific tasks, and/or assignments to use while they are observing. If multiple PSTs are observing, each can be assigned an individual or small group of students on which to focus. It can be tempting for PSTs to roam the classroom and see what every student is doing but focusing on just one or a few students can help discern specific supports or barriers to their learning. It may also be helpful for PSTs to have specifically identified points on the research lesson plan where it will be useful to collect data and for which something is going on that data can be collected, such as open-ended questions, discussion, or student presentations (Murata, 2021).

The Role of Knowledgeable Other: Enhancing Educators' Conversations

Lewis and Hurd (2011) suggest that a potential member of a lesson study team is a "knowledgeable other" (Choy et al., 2021). Specifically, their role "is to raise questions, add new perspectives, and to be co-researcher, not to tell others what to do" (Lewis & Hurd, 2011, p. 33). Researchers recommend guidance from a knowledgeable other to improve PSTs experience and learning during the lesson study cycle (Chassels & Melville, 2009; Michaels, 2020A, 2020B; Myers, 2013; Parks, 2008). Typically, in PST education, TEs are the knowledgeable other. They facilitate a pre-lesson discussion with classroom teachers and PSTs before the lesson is taught. They also provide guiding questions and prompts to help PSTs know how to collect data on student learning during the research lesson. They help ensure understanding of lesson study procedures and the research lesson's learning goals for K-12 students. Other people who may serve in this role can be a curriculum director, subject-specific teacher leader, content professor, or an in-service classroom teacher. The lesson study group must decide what the role of the knowledgeable other is meant to be and how this person can help them best prepare for the *teach* step and reflect on student observations afterward. This person may contribute ideas about the direction of the lesson, suggest high-quality curriculum or other useful materials, facilitate reflection, access relevant subject matter knowledge and pedagogical strategies, and help prevent PSTs from making their own subject-matter mistakes or following their own misconceptions by challenging them to think through problems and content in new ways.

Implementation of the *Teach* Step: Facilitating the Development of PSTs' Pedagogical Knowledge

The lesson(s) that make up the *teach* step may be organized in several different ways. Sometimes, lesson study involves one research lesson that is implemented more than once but with a different group of students, with each iteration of the lesson becoming better aligned with student needs and understanding as the observers collect data on the participating students and update the research lesson accordingly (Lewis & Hurd, 2011). In this format, the teacher may change for each implementation of the lesson as the lesson study team moves to different classrooms (LSMSTEC, 2021 [Hummer poster]). This structure may be difficult given the number of classrooms available, their varying grade levels, and considering that the TE and PSTs are guests within a local school. PSTs may have the opportunity to teach just one lesson in their assigned classroom and collect data on student learning. Even with this limited opportunity, it will help PSTs understand what professional collaboration around developing mathematics and science lessons may look like as they work with their classroom teacher on creating a lesson that follows specific objectives to meet identified student needs. Conducting this one lesson can also provide a context to reflect on student learning for additional lessons that may or may not be part of a lesson study process.

Given the constraints inherent in teacher preparation programs, we suggest two possible structures as productive implementation opportunities that the authors of this chapter use for the *teach* step when working with PSTs in K-12 school settings. These structures are gradual; PSTs participate in the first example early in their teacher preparation, and as they move through the program, they are ready for the second.

1 A K-12 classroom teacher may implement a lesson with their students, while PSTs are observers who collect data on student learning and teacher pedagogy (LSMSTEC, 2021 [Michaels poster]).
2 A unit can be implemented by PSTs, resulting in a series of lessons that build from one to the next using the same learning standard or concept. In this format, the data collection during lesson implementation informs plans for the next unit lesson (Dudley & Lang, 2021; Fujii, 2014; LSMSTEC, 2021 [Glen poster, Suh poster]).

Below, we provide additional details for these two possible ways a TE might implement lesson study with PSTs.

Teacher of Research Lesson

Classroom Teachers Teach Research Lesson

A growing body of research indicates that university-school lesson study experiences should purposefully connect TEs and PSTs with K-12 teachers via the methods courses within a teacher preparation program. The lesson study experience in methods courses can include classroom instruction, thereby providing high-quality learning experiences for PSTs (Burroughs & Luebeck, 2010; Cajkler et al., 2013; Michaels, 2015; Michaels, 2020A, 2020B). When TEs and K-12 teachers are a collaborative team in the lesson study process, PSTs are provided opportunities to learn to teach through guided, structured experiences that connect coursework theory to teaching practice in schools. Specifically, a common focus on pedagogical goals, modeled by classroom teachers and facilitated by TEs, provides important links between theory and teaching practice for PSTs.

One lesson study program for PSTs is in the San Francisco Bay Area at Dominican University of California. As the TE of the mathematics and science methods courses, first author Rosemarie Michaels collaborates with individual classroom teachers on complete lesson study cycles. She structures and facilitates the lesson study sessions, taking on the role of knowledgeable other. Classroom teachers in elementary schools are chosen to lead based on their expertise in mathematics or science pedagogy and ability to model inclusive, effective teaching strategies to engage *all* students in the learning process, including multilingual learners (Michaels, 2020B). PSTs participating in this structure of lesson study are novices, specifically, freshmen through juniors or beginning graduate students. As such, they are not yet developmentally ready to lead lesson study experiences in K-12 classrooms. In order to prepare them to do so, the TE and classroom teachers collaborate on the research lesson that is subsequently taught by the classroom teacher, while PSTs are observers and participants in this process. As PSTs progress through the teacher preparation program, they gradually take on more responsibility during lesson study, culminating in formal lesson study experiences as the lead teacher within a collaborative group of PSTs (as described by the second author in the next section below).

The protocol for a lesson study cycle in Michaels' mathematics methods course follows. A classroom teacher and Michaels discuss the lesson and determine specific pedagogical strategies and academic goals based on California's Standards for Mathematical Practice for K-12 students and PSTs (California Commission on Teacher Credentialing, 2016), and that will be modeled during the lesson (e.g., make sense of problems and persevere in solving them, construct viable arguments, and critique the reasoning of others). The classroom teacher plans the research lesson to address their students' academic needs and learning goals. Michaels then works with PSTs to analyze the lesson plan during a meeting of her course on campus. Particular attention is paid to the student learning goals, the teaching strategies planned, and the data PSTs will collect during the research lesson. They review the focus Standards for Mathematical Practice and strategies the teacher plans to use to engage multilingual learners during the lesson and discuss the rationale for their implementation. In addition, Michaels previews the observation form which will be used as a data collection tool (Michaels, 2020B).

On the lesson study day, the classroom teacher, PSTs, and Michaels meet at the school site before the lesson. As the knowledgeable other, Michaels leads the pre-observation meeting, wherein the classroom teacher discusses the planning process, rationale for the use of instructional strategies, and goals for the lesson. PSTs ask clarifying questions to better understand the teacher's planning process and "teacher thinking." In addition, the classroom teacher discusses how PSTs can participate with students during the lesson. For example, the classroom teacher may ask PSTs to participate by working with a small group of students at a specific time during the lesson or to focus on one student who needs extra support. Finally, Michaels reviews the goals for PSTs and data collection procedures. Data is collected in two areas: Student learning of the lesson's goals and effective teaching strategies.

The classroom teacher teaches the lesson to their students. PSTs observe and take notes on a data collection document, which has space to record:

1 strategies the teacher uses to engage students in two Standards for Mathematical Practice, for example:

   1 model with mathematics,
   2 construct viable arguments;

2 strategies the teacher uses to engage and include multilingual learners;
3 what the students are learning and evidence to support this claim; and
4 questions for the *reflect* step.

While the teacher is teaching the lesson, Michaels guides PSTs to notice best practice by quietly highlighting important aspects of the classroom teacher's pedagogy and pointing out aspects of student learning or misconceptions. PSTs typically participate in guiding a small group or individual students for part of the lesson; previous work suggests this enhances PSTs' learning while the TE guides the experience (Michaels, 2020A, 2020B). Immediately following the lesson, all participants meet to debrief and reflect on the lesson and learnings. During the next methods class, PSTs continue to analyze the lesson and student learning or misconceptions.

Some TEs from the LSMSTEC (2021) use lesson study in similar ways—where the classroom teacher conducts a lesson with their own students while PSTs observe. Komatsubara (LSMSTEC, 2021), a TE at High Tech High Graduate School of Education, collaborates with classroom teachers to facilitate PSTs' understanding and implementation of anti-racist pedagogies during the student teaching stage of their teacher preparation program. PSTs plan the lesson and then teach it to their peers during a methods course class meeting. PSTs provide feedback to their peers on the lesson plan and instruction. Then the formal lesson study session is held at a secondary school site; the classroom teacher teaches the lesson, modeling the best pedagogy in mathematics or science, including anti-racist pedagogies, like Dominican University's program. PSTs closely observe focal students and collect data on the students' thinking but do not teach the lesson.

PSTs Teach Research Lesson

Field experiences during mathematics and science methods courses lend themselves to a collaborative and dynamic experience for PSTs to engage in lesson study as a team with a classroom teacher in a school. One program like this is in southeast Massachusetts at Bridgewater State University with TE and second author Nicole Glen. Guided by her as the knowledgeable other, a pair of PSTs plan a series of science lessons that make up a standards-based unit of instruction. They alternate the teaching of their lessons across the span of the unit. To engage in lesson study, when one PST is the lead teacher, the other is the observer, collecting data on the students during the lesson. After teaching, Glen holds a debriefing and reflection session that is led by the PST-observer to discern struggles and successes of the students and make appropriate adjustments to the next lesson in the unit sequence, which they will lead. The PST who was the observer is then the lead teacher for the subsequent lesson while the PST who previously led the lesson becomes the observer. These roles switch back and forth for four lessons (each PST is the lead teacher for two lessons and the observer for two lessons).

The role of the observer may be one of the most difficult for PSTs to take on. One reason for this is that they desperately want to help their peer PST; they do this by passing out materials and papers, attending to classroom management, and more. Being the observer, therefore, has several goals. One is that it helps the lead PST learn to manage the classroom on their own, and because the PSTs in this program move onto student teaching immediately following the semester of this experience, this is good practice for them. The second goal is that it allows the observing PST to focus on the students. By providing PSTs with foundational knowledge and practice observing and attending to student learning needs, it can set them up for future success in their pedagogical growth and decision-making skills for teaching (Blömeke et al., 2015). This also helps build PSTs' "habit and skill of careful observation" of student learning (Lewis & Hurd, 2011 p. 26).

In the university classroom, Glen models and PSTs practice the observer role prior to the unit of instruction taking place. They do this by watching videos of student talk at different points of a lesson sequence (e.g., beginning, middle, and end of a science lesson; or within parts of the engineering design process). PSTs take detailed notes on and share their observations of student talk using a template that Glen created based on research about conceptions of young children in science

TABLE 7.1 Template for Data Collection on Student Conceptions

Lesson date	*Lesson topic*
Partial understandings	*Full understandings*
(Record here specific quotes; how many students, on average, are thinking that; who said/did what and when)	(Record here specific quotes; how many students, on average, are thinking that; who said/did what and when)
Alternative understandings	*Everyday language and/or experiences of students*
(Record here specific quotes; how many students, on average, are thinking that; who said/did what and when)	(Record here specific quotes; how many students, on average, are thinking that; who said/did what and when)
Attitudes and/or behaviors of students	
(Record here how students seem to feel about what they are doing, how you know they feel this way; what you are learning about particular students in the class or the class as a whole; how they are acting and/or reacting to what they are working on)	

(Windschitl et al., 2018). An example of this is in Table 7.1. Glen also guides PSTs to look at sample student work with the same categories in mind.

There are several complementary ways that a TE might engage PSTs in this type of lesson study within a K-12 school. One example is TE Suh from LSMSTEC (2021) and the elementary education PSTs from George Mason University. Here, teams of PSTs plan and teach a three-to-five lesson project-based learning unit that incorporates mathematics and science in a local multilingual school. They co-teach their lessons, which are video recorded to observe and annotate student concepts and ideas afterward. A second example is TE Hummer from LSMSTEC (2021) and the secondary education mathematics and science PSTs at West Chester University of Pennsylvania. Teams of three to four PSTs collaboratively plan lessons during their methods course. One PST from the team teaches the lesson in a local classroom, and the other team members leave their classrooms to observe the lesson. This cycle continues until all team members have taught the planned lesson.

A final example is TE Reins from LSMSTEC (2021) and the secondary education mathematics PSTs from the University of South Dakota. In this model, the classroom teachers provide topics and PSTs work in teams to create a lesson for that topic. As part of the *study* step, PSTs interview students in the host classroom to learn more about their thinking. PSTs then *plan* and *teach* the lesson and the rest of the methods course PSTs attend the lesson to collect data on student thinking and learning, using a provided template and discussion protocol. The lessons are video recorded from multiple angles so that the observing PSTs and knowledgeable others, which includes the TE and an invited mathematics professor, can review the recordings after the lesson to gain more detailed insight on student thinking.

Recommendations for Teacher Educators

Researchers recommend that lesson study with PSTs be thoughtfully structured and facilitated by a knowledgeable other, such as the TEs of methods courses (Chassels & Melville, 2009; Michaels, 2020B; Myers, 2013; Parks, 2008). The goal should be to provide PSTs with authentic teaching experiences that also examine student learning through careful observations by PSTs during the *teach* step. Lewis and Hurd (2011) recommend TEs be mindful of their lesson study practice to ensure that the research lesson is not just a demonstration lesson by an expert teacher with no room for interrogation of the lesson effectiveness or analysis of actual student learning.

An equally important component of the *teach* step is a connection with local K-12 schools and classroom teachers eager to work with TEs and PSTs in this way. We have found that having classroom teachers who allow PSTs to practice mathematics and science teaching using thoughtfully planned research lessons has led to effective learning experiences for all involved. The classroom teachers appreciate the time they can more carefully observe specific students during a lesson that PSTs implement. They enjoy showcasing their strengths and improving their development as teachers by thinking intentionally about and using key instructional strategies when PSTs are observing them.

When working with K-12 schools and classroom teachers, several points may need to be negotiated ahead of lesson study. For example, if it is expected that the classroom teacher will debrief with PSTs, providing them with specific facilitation guidelines or prompts that support the TE's lesson study goals is important so they do not only focus on teaching strategies to the exclusion of student learning. Finding classroom teachers who are knowledgeable others about mathematics and science content and pedagogy can help further PSTs' learning about reform-oriented mathematics and science instruction. It may help to work with the principal of the school in which PSTs are enacting lesson study to negotiate release time for the classroom teachers to help plan the lesson ahead of the *teach* step and debrief with PSTs after (LSMSTEC, 2021 [Glen poster]). An alternate idea for a school collaboration is to plan for PSTs to move into other classrooms in the school for a short time, or for the school to host PSTs who are not teaching at the school to be visitors for a day (LSMSTEC, 2021 [Hummer poster]) to be observers during the *teach* step enacted there. Finally, for methodology courses that do not follow a typical academic year, daytime structure, finding or creating teaching experiences for PSTs such as after-school and summer programs (LSMSTEC, 2021 [Glen poster, Suh poster]) can be equally useful for facilitating lesson study.

An effective lesson study program with PSTs should engage them in a variety of lesson study experiences over the course of their teacher preparation, beginning in introductory courses and continuing in more advanced forms through upper-level methodology courses. PSTs can gradually take on responsibility for collaboratively planning and teaching research lessons, culminating in a full lesson study cycle with peers and classroom teachers in a K-12 school setting during student teaching. The lesson study program facilitated by Michaels at Dominican University of California provides a good example of this type of progression, which is structured by TEs of methods courses. Beginning PSTs participate in classroom teacher-led lesson study, as described earlier. The classroom teacher conducts the *plan* and *teach* steps of the research lesson while PSTs observe and collect data. PSTs also participate with the TE and classroom teacher during the *reflect* step of lesson study. As they advance through their degree program, PSTs collaborate with peers to *plan* and *teach* lesson study experiences to small groups or a whole class of students during fieldwork in schools and guided by TEs, as described earlier. They may continue to participate in classroom teacher-led lesson studies. As a culminating experience during student teaching, PSTs are actively involved in a traditional lesson study session. They collaborate with a classroom teacher and the TE of their student teaching seminar to *plan* a research lesson over a 2–3-week period. Then, one PST *teaches* the lesson while their peers, TE, and classroom teacher observe and collect data on student learning, which is followed by the *reflect* step. Then, the lesson is revised and retaught by a different PST in a new classroom. The success of this lesson study program has been described in more detail (see Michaels, 2020A). However, it may be improved further by inserting a step recommended by Post and Varoz (2008): A classroom teacher teaches the first lesson, while a PST teaches the revised lesson.

Concluding Thoughts

The *teach* step of lesson study provides PSTs with the opportunity to delve deeply into the craft of teaching and learning. Guided by TEs and aided by K-12 classroom teachers, PSTs learn to carefully observe classroom instruction, connecting pedagogical theory presented during methods courses and actual teaching practice. Simultaneously, PSTs are guided to develop the skills needed to assess student learning by observing and collecting data during classroom instruction. In this way, we are preparing future teachers through the shared experience of looking at teaching with a critical eye while preparing for reflective conversations during the next step of lesson study.

References

Blömeke, S., Hoth, J., Döhrmann, M., Busse, A., Kaiser, G., & König, J. (2015). Teacher change during induction: Development of beginning primary teachers' knowledge, beliefs, and performance. *International Journal of Science & Mathematics Education*, *13*(2), 287–308.

Burroughs, E. A., & Luebeck, J. L. (2010). Pre-service teachers in mathematics lesson study. *The Mathematics Enthusiast*, *7*(2), 391–400.

Cajkler, W., Wood, P., Norton, J., & Pedder, D. (2013). Lesson study: Towards a collaborative approach to learning in initial teacher education? *Cambridge Journal of Education, 43*(4), 537–554.

California Commission on Teacher Credentialing. (2016). California teacher performance expectations. https://www.ctc.ca.gov/docs/default-source/educator-prep/standards/adopted-tpes-2016.pdf?sfvrsn=0

Chassels, C., & Melville, W. (2009). Collaborative, reflective, and iterative Japanese lesson study in an initial teacher education program: Benefits and challenges. *Canadian Journal of Education, 32*(4), 734–763.

Choy, B. H., & Lee, C. K. E. (2021). 5. Going deeper into lesson study through *kyouzai kenkyuu*. In Murata (Ed.), *Stepping up lesson study: An educator's guide to deeper learning* (pp. 39–51). WALS-Routledge Lesson Study Series.

Dudley, P., & Lang, J. (2021). 3. How case pupils, pupil interviews and sequenced research lessons can strengthen teacher insights in *how* to improve learning for all pupils. In A. Murata, & K. E. Lee (Eds.), *Stepping up lesson study: An educator's guide to deeper learning* (pp. 14–26). WALS-Routledge Lesson Study Series.

Fujii, T. (2014). Implementing Japanese lesson study in foreign countries: Misconceptions revealed. *Mathematics Teacher Education and Development, 16*(1), 2–18.

Lewis, C., Friedkin, S., Emerson, K., Henn, L., & Goldsmith, L. (2019). How does lesson study work? Toward a theory of lesson study process and impact. In R. Huang, A. Takahashi, J. daPonte (Eds.), *Theory and practice of lesson study in mathematics.* (pp. 13–37). Springer. https://doi.org/10.1007/978-3-030-04031-4_2

Lewis, C. C., & Hurd, J. (2011). *Lesson study step by step: How teacher learning communities improve instruction.* Heinemann.

LSMSTEC (Lesson Study for Mathematics and Science Teacher Educators Conference. (2021). Online.

Michaels, R. (2015). Bringing lesson study to teacher education: Simultaneously impacting pre-service and classroom teachers. *Journal of Scholastic Inquiry: Education, 4*(1), 46–73.

Michaels, R. (2020A). Lesson study with preservice teachers: Learning to teach English language learners. *Journal of Scholastic Inquiry: Special Edition, 10*, 87–110.

Michaels, R. (2020B). Mathematics lesson study in elementary preservice teacher preparation. *Journal of Scholastic Inquiry: Special Edition, 10*, 111–133.

Murata, A. (2021). 2. Lesson study as research: Relating lesson goals, activities and data collection. In A. Murata, & K. E. Lee (Eds.), *Stepping up lesson study: An educator's guide to deeper learning* (pp. 4–13). WALS-Routledge Lesson Study Series.

Myers, J. (2013). Creating reflective practitioners with pre-service lesson study. *International Journal of Pedagogies and Learning, 8*(1), 1–9.

Parks, A. N. (2008). Messy learning: Pre-service teachers' lesson-study conversations about mathematics and students. *Teacher and Teacher Education, 24*(5), 1200–1216.

Post, G., & Varoz, S. (2008). Lesson study groups with prospective and practicing teachers. *Teaching Children Mathematics, 14*(8), 472–478.

The Lesson Study Group at Mills College. (2022). *Teach.* Retrieved from https://lessonresearch.net/conduct-a-cycle/teach/

Windschitl, M., Thompson, J., & Braaten, M. (2018). *Ambitious science teaching.* Harvard Education Press.

8

THE REFLECT STEP

Consolidating Learning and Setting Future Goals

Sharon Dotger and Kelly Chandler-Olcott

Reflect is the last formal step within a lesson study cycle, and in contrast to the other steps within a lesson study cycle, can be completed in 1–2 hours. As Lewis and colleagues (2019) have argued, the *reflect* step has three primary goals. The first, "articulate what individual team members and observers learned from the lesson study cycle, so that this knowledge becomes available to others within and outside the team" (p. 29), often takes place with a structured post-lesson discussion. The second, "individual participants integrate knowledge from the lesson study cycle into their own thinking and practice" (p. 29), may occur during the post-lesson discussion, but ideally this goal is met more fully over time, as the lesson study participants engage in the daily work of teaching within their classrooms and work with colleagues within future lesson study cycles. The third, "for educators to strengthen their commitment to improvement of their own knowledge and practice and that of colleagues" (p. 29), is an articulation of a learning stance that overlaps with existing literature that addresses how groups of teachers can work together to advance instructional improvement (Bryk et al., 2015).

In this chapter, we will address how these goals apply to preservice teacher (PST) education and how they connect with the other steps of a lesson study cycle. In doing so, we will summarize colleagues' contributions about *reflect* during the Lesson Study for Mathematics and Science Teacher Educators Conference (LSM-STEC) and draw on examples from our own work using lesson study cycles with PSTs in two different pathways to certification.

Applying the Goals of *Reflect* in PST Education

As part of their preparation programs, PSTs are regularly required to reflect on their practice, as a long-valued key element of effective teacher development (q.v. Dewey, 1933). Yet, a persistent challenge in teaching PSTs to reflect is helping

DOI: 10.4324/9781003326434-11

them conceptualize problems of practice, while resisting overly prescriptive instructions for reflection and avoiding reflections that are insufficiently detailed (Clará et al., 2019). Complicating matters, there are a multitude of interpretations across the field of teacher education of what it means to reflect (Korthagen, 2001). Lesson study has mechanisms to address both concerns within the *reflect* step. Illustrating these mechanisms requires a brief look across the lesson study cycle at what is available to reflect upon.

Beginning with the *prepare* step, PSTs develop norms for cooperation. As the teacher educator helps form research teams with cooperating teachers through field placements, they may negotiate with the cooperating teachers, school leaders, and field supervisors to design the experience. Depending on the composition of the lesson study team, these professionals may be on the team, observers in research lessons, or interested audience members for the learning the PSTs will articulate within the *reflect* step. Therefore, *prepare* sets up one avenue for reflection, cooperation with others, and creates an audience beyond the teacher educator with whom the PSTs could share their learning, an important outcome of the *reflect* step.

During the *study* step, PSTs articulate a research theme and build knowledge of the content topic. Along the way, they study content, standards, research articles, and instructional materials. These investigations influence their work in the *plan* step, where the work from *study* manifests in a detailed lesson plan designed to improve a specific teaching and learning problem. Across each of these steps, PSTs design with intention. These designs, and their intentions, set up additional topics for discussion and evaluation during the *reflect* step.

Within the *teach* step, the members of the lesson study team see how ideas from *prepare, study*, and *plan* come together when the lesson is taught live to students. The taught lesson is thus a test of the ideas established in the steps that led up to it. Observers of the research lesson, which hopefully include PSTs, also can observe student thinking in live time, an important skill to develop when aiming for responsive teaching.

The *reflect* step is punctuated by the post-lesson discussion, an essential element of the step. Its purpose is to build shared understanding among the lesson study team members and the observers about the student thinking that emerged during the lesson, how the design of the lesson influenced that thinking, and to build norms for discussing teaching that build a shared view of instructional practices that influence student learning. Traditionally, post-lesson discussions, which initiate the *reflect* step, follow the *teach* step almost immediately. The timing of the post-lesson discussion is close to the research lesson to capitalize on the fact that a group of professionals have gathered to study teaching, and gathering again later is often difficult logistically. Further, discussing the learning from the lesson almost immediately can galvanize the takeaways, help teachers learn from the observations of their colleagues, and spur further lesson study cycles.

Lesson study scholars have developed protocols for post-lesson discussions (see, for example, those available at lessonresearch.net) that proceed with opening comments from the teacher of the lesson. Next, team members who have been involved in the other steps of the lesson study process offer their observations, followed by observations from other observers, if there are any. Lesson study teams are often supported by a knowledgeable other, which in the in-service context may be a teacher educator, an instructional coach, or other leader in the school. During the post-lesson discussion, this knowledgeable other may also offer a few observations. When research lessons are public (and thus involve observers in addition to the lesson study team members), the post-lesson discussion often closes with insightful statements by a final commentator. Final commentators should have deep knowledge of lesson study and the content taught in the lesson, to be able to offer the lesson study team a few select suggestions for their future lesson study cycles.

The *reflect* step often concludes with one final meeting of the lesson study team, where they discuss their learning across the cycle and set goals for their future work. This final meeting may be attended by the knowledgeable other who supported them throughout the lesson study cycle.

In our experiences with lesson study, the *teach* and *reflect* steps are connected to one another in time and space. In the sections that follow, we overview trends in how our LSMSTEC colleagues presented their *reflect* step practices. After this overview, we describe two ways we have organized the *reflect* step with our PST candidates. To do so, we describe the programmatic context that surrounds our design of the *reflect* step and details the decisions we make as teacher educators as we have iterated our structures for *reflect* overtime.

Trends in *Reflect* in Teacher Education at LSMSTEC

As our LSMSTEC colleagues discussed their practices during *reflect*, their ideas coalesced around helping PSTs answer a pair of central questions: Is the lesson doing what we want it to do? If not, what do we need to change for next time? Through further discussion, LSMSTEC participants noted that *reflect* worked well when PSTs could teach, observe, and reflect on the research lessons with children in schools. Some teacher educators at LSMSTEC reported that when research lessons are taught in a micro-teaching setting, sometimes the PSTs believe the lessons would have gone differently if they could have been taught with "real kids," perhaps undermining the PSTs' learning outcomes. Questions emerged among the participants regarding the relationship between team-based discussion, like the post-lesson discussion in the lesson study cycle, and the individual reflection, often demonstrated through writing, that is a common feature of many teacher education programs.

A theme from the discussion at LSMSTEC was focused on teacher educators' hopes that by PSTs conducting lesson study in K-12 schools during their field placements or student teaching experiences, the deep reflection on the lesson and its learning outcomes might spur host teachers to want to do lesson study even when the PSTs were no longer in the school. Alternatively, teacher educators also hoped that by experiencing lesson study in the teacher education program, PSTs might be more inclined to start lesson study in their future schools or join teams if lesson study was already occurring.

To support PSTs' action during *reflect*, teacher educators deployed various strategies. For example, a PST observing a peer teach the research lesson shared out specific quotations they heard from students and detailed descriptions of what they saw students do. Afterward, the observing PST and the teaching PST worked together to determine if subsequent lessons in the unit could be taught as planned or needed adjustment (see Glen poster). This practice was similar among multiple LSMSTEC participants, with some emphasizing the importance of the *reflect* step occurring immediately after the conclusion of the lesson (i.e., Hummer poster). In another case, the teacher educator realized that when PSTs observe research lessons that take place in college classrooms, they may need a primer about the purpose of the lesson, especially if the lesson design diverges from their previous collegiate mathematics learning experiences (see Graham poster). Some teacher educators encouraged PSTs to create an artifact that they shared with their classmates intended to help the PSTs remember their learning in future teaching (see Komatsubara, Taylor, & Sharrock posters).

These examples demonstrate both some commonalities and some variation across conference participants in their approach to the reflect step. In the following section, we share more extended examples of how we orchestrate this step in our work, to make the linkages to other steps clearer as well as explore the potential effects of such similarities and differences.

Case 1: Undergraduate Elementary Teacher Education

Sharon teaches a K-6 science curriculum and methods course to PSTs seeking both general and special education certifications. PSTs enroll in this course during the semester immediately before they student teach. The science methods course is accompanied by methods courses in mathematics and intermediate grades literacy. Together, these three courses have assignments the PSTs complete in a 6-week full-day placement in local schools.

Sharon has used lesson study principles for two iterations of the course, differentiated by the interest and availability of local schools to host PSTs as a whole group or not. When local schools would like to host the whole group (Case 1.1), PSTs form a lesson study team for each class in the grade level—so if there are five classes of fifth graders in the school, we form five lesson study teams. In one iteration, PSTs planned ten consecutive 30-minute science lessons focused on evaluating the relationship between the sun's changing position in the sky and the length and direction of shadows that result. Prior to the research lesson, teams study and plan together, then rehearse their lessons with each other before they teach students. Rehearsals often include mock-ups of *bansho* (board work) (Figure 8.1), some of which involve modifying lesson plans from their designs in available instructional

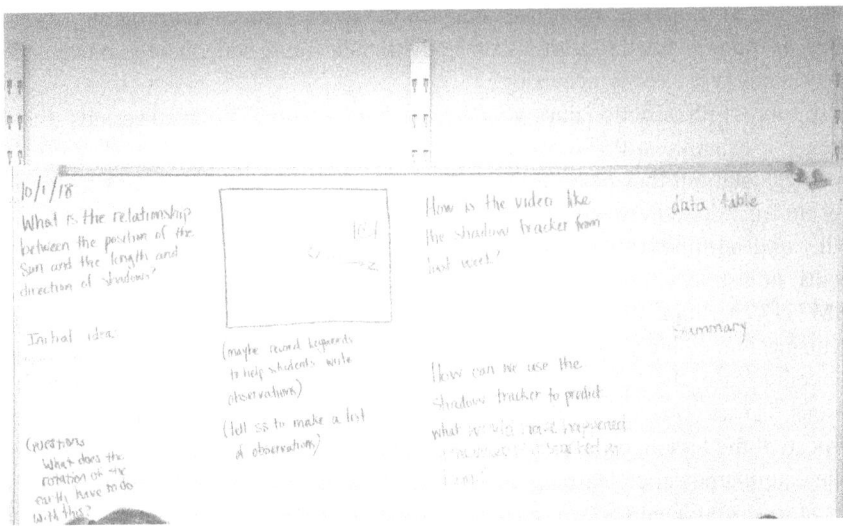

FIGURE 8.1 Example of a *Bansho* Mock-Up.

materials. For 2 weeks, PSTs take turns teaching research lessons to the fifth graders. They are observed by teammates, the course instructor, the host teachers, and graduate students learning to become teacher educators. Post-lesson discussions are held immediately afterward in an empty classroom within the school.

In this case, since five PSTs taught a lesson in separate classrooms that they co-planned and rehearsed together, and since the remaining PSTs observed those lessons and the lessons built on one another and into a broader understanding across the unit, our post-lesson discussions were whole group. Building from protocols for a post-lesson discussion (see lessonresearch.net for examples), we began the post-lesson discussion with the five teachers who taught the research lesson sharing their observations of student thinking from each iteration of implementation. Sharon recorded these observations from the lesson teachers on the board and augmented them with additional data from the PSTs who watched the students as the lesson unfolded. After sufficient understanding of what students were thinking in the lesson was established and similarities and differences between enactments and learning outcomes were represented on the board, the conversation turned to how these observations could inform the design of the next day's instruction. Although we studied the unit trajectory prior to the initiation of the first lesson, modifications were needed as we learned more about students' thinking. When the post-lesson discussion concluded, groups photographed the board and worked in their classroom-based teams to design and rehearse the lesson for the next day. In this design, the post-lesson discussion yielded ideas that fed right back into a *plan/teach/reflect* cycle for the remainder of the unit. For example, in a subsequent lesson about the causes of day and night, PSTs focused on eliciting students' initial explanations for the phenomenon prior to launching a hands-on investigation.

At the conclusion of this 2-week experience, a final whole group discussion was held, followed by individual writing guided by prompts. The whole group discussion invited the PSTs to talk about what they learned about children's scientific thinking, instructional methods for eliciting that thinking, and questions that remained or emerged. The writing prompts asked them to think about these same issues for themselves and set goals for their own future practice as they transitioned first to student teaching and then to their career. Although the writing prompts could have been completed outside of class, we asked the PSTs to complete them as soon as the whole group discussion ended so that we might capture their ideas before they had the chance to fade away.

When local schools cannot host the class (case 1.2), PSTs form teams on-campus prior to the start of their placement. If there are 15 PSTs in the class, three groups of five PSTs are formed. Each group is given a science topic with accompanying standards and instructional materials. Using microteaching, Group A teaches their lesson to peers in groups B and C, rotating every hour. Each lesson is 30 minutes long, followed by a post-lesson discussion among Group A and the teacher educator. In this type of implementation, whole class planning and rehearsal is not reasonable, as each team is working on a different science topic. Whole class planning and rehearsal would reveal the lesson to classmates who need to be students and therefore, the enactment of the lesson would require playacting by the PSTs who already know what would happen and yet be in a student role. Our teaching experience suggests that PSTs do not necessarily already know the content of ambitious elementary science units and experiencing it first-hand and without prior awareness of the flow of the lesson leads to a more authentic approximation for the lesson teachers.

Thus, each on-campus class meeting of 3 hours is divided into thirds. In the first third, team A (or whichever team goes first) has a member teach the lesson, other teammates observe. As soon as the lesson ends, team A vacates the classroom. This permits team B time to reset the classroom for their lesson. While team B is setting up their lesson, team A has a short (20 minute) post-lesson discussion facilitated by the teacher educator. Protocols like those in Case 1.1 are used but may be shortened in time since team B will teach at the top of the hour and because there are fewer teachers and observers to share their observations of the lesson. The teams rotate through this process, hour-by-hour, class-by-class, until all PSTs have had the opportunity to teach a research lesson to their peers.

Case 2: Graduate Level Initial Teacher Certification

Case 2 was embedded within a research-practice partnership (Penuel & Gallagher, 2017), that is associated with the second of 11 courses in a graduate-level teacher education program. The students enrolled in the program have undergraduate degrees in the content area they will teach, but little to no prior teaching experience. This course, taught by Kelly, focused on literacy across the curriculum, was taught in the summer, and prepared the PSTs to design and enact literacy-infused week-long units to local youth who register for a half-day summer enrichment program (Chandler-Olcott et al., 2018, 2021). The PSTs worked together in teams to teach small groups of youths and each team of PSTs was supported by a mentor teacher hired from the local school district, as well as by Kelly or another teacher educator teaching the literacy course. Sharon participated in the project as a lesson-study consultant, often serving as the outside commentator for the research lessons.

The summer program, called Camp Questions, was 3 weeks long for the youth. One week prior to the start of the camp, Kelly and the PSTs studied the principles of disciplinary literacy as they apply to multiple school disciplines, including but not limited to science. Kelly, based on previous consultation with youth and teachers from the school district, designed a 4-day unit on a topic of local interest that all PSTs taught during the first week of camp. In the second and third weeks for the youth, PSTs planned additional lessons supported by unit organizers and the wisdom of their mentor teachers. The units were designed so that on the first day of the 4-day unit, students articulated questions about topics such as how to improve school lunches or develop joint programs between youth and community elders. With these questions in mind, on the second day they crafted surveys and interview questions to gather perspectives about these topics from their peers. On the third day, they analyzed the data they gather to formulate answers to their questions. On the fourth day, they designed representations to share their ideas with an outside audience, like the mayor or the school superintendent.

Given this unit arc, we planned the third lesson in the 4-day arc as a research lesson. Teacher educators formed interdisciplinary teams of two or three, supported by a designated mentor from the school district. We selected these days as research lessons because we believe the nature of the task—making sense of data and formulating answers—presents were good opportunities to make students' thinking public and would therefore generate plenty of data for observers to gather. To strengthen the partnership between our university and the local school district, we invited local teachers, school leaders, and university faculty to attend the research lessons and participate in the post-lesson discussion.

Our design of the post-lesson discussion has undergone multiple iterations over the 6 years of the Camp Questions project. In all iterations, lesson study cycle steps were infused with literacy principles, and these were evident in the post-lesson discussion and in the final projects the PSTs authored by the end of the course. In our first year of implementation, when the research lesson ended, the students left the room, and the classroom seating was reorganized to accommodate a structured discussion between the members of the teaching team and the observers of the research lesson. Yet, despite our instructions, we noticed the PSTs struggled to comment from their observational notes. Instead, their comments were more oriented to evaluation of teacher performance, a dicey situation given the novice status of the PSTs and everyone's uncertainty about norms for publicly discussing a research lesson.

Our use of the same room for post-lesson discussion was initially a product of our tight schedule. Therefore, we made a few adjustments. We scheduled an hour for the post-lesson discussion, down the hall and in a separate room. We used the first 5 minutes or so to encourage the observers to read over their notes; flag moments that seemed significant, especially relative to the research theme and learning objectives articulated in the lesson plan; and allow the teacher of the research lesson a few moments to record their own thoughts. The elongated time for the post-lesson discussion allowed for more coaching of novices on how to conduct a post-lesson discussion and allowed time for comments to be identified intentionally. Therefore, in future iterations, we developed a structured mechanism that asked observers to use small, colorful sticky notes to flag their observations.

Additionally, we posted the lesson objectives on the board in the post-lesson discussion space and asked the observers to make connections between the flagged moments in their notes and the lesson objectives.

Over time, these tweaks to our procedures led to PSTs' greater focus on evidence of student learning, rather than teacher performance, and noticeable reduction in their anxiety about being observed—and thus potentially judged—by others. The adjustments allowed us to home in during the post-lesson discussion on the literacy objectives for the lesson, as opposed to general instructional delivery, and thus made the experience better aligned with the content for the course in which PSTs were enrolled. Over time, the refinements also helped our mentor–teacher partners, who tended to return to the program year after year, develop their own observational and reflective skills, which enhanced their ability to coach and model professional behaviors for our candidates.

Challenges in Adapting *Reflect* to the Pre-Service Context

As noted by Lewis et al. (2019), one purpose of the *reflect* step is to share learning with an outside audience. Commonly, the assignments PSTs complete in teacher education are written for university faculty, field supervisors, or host teachers to read. They are rarely disseminated beyond these individuals. Additionally, in the cases we discuss herein, lesson study is not yet an activity conducted by classroom teachers as a regular component of their professional development. It has occurred, but by individual teams of teachers conducting isolated cycles. In other words, lesson study is not yet school-wide. In a school-wide lesson study context, other teachers who are conducting lesson study are an audience for lesson study teams to communicate with. We imagine that in the future, PSTs who have experienced lesson study in their teacher education programs may be well positioned to contribute to the discourse on lesson study with in-service teachers after they enter the profession.

Additionally, it is currently unlikely that the PSTs who have had these lesson study experiences will have additional opportunities to participate in lesson study. The *reflect* step should generate new learning for the participating teachers and generate new questions. Without a continued opportunity to conduct lesson study, there are not easily identifiable mechanisms to support additional inquiry into their practice or to do it with colleagues. Lewis et al. (2019) pointed out that *reflect* has the potential to re-orient teachers to developing positive dispositions for working collaboratively and learning to improve their practice across their career—yet these opportunities need to be continuous.

Concluding Thoughts

We argue that the *reflect* step of lesson study can address concerns about teacher reflection more broadly. Since PSTs began conceptualizing a problem of practice they could address in the *prepare* step, discussing that problem during the *reflect* step allows them to see the problem from first articulation all the way through the cycle, ending with observations and study of student thinking. The post-lesson discussion was a venue through which the PSTs evaluated their conceptualizations of the problem, drawing on the data they gathered from observing the live research lesson. Grounding additional writing, something we both do in our implementation of lesson study with PSTs, in the specific evidence gathered during the *teach* step and discussed in the *reflect* step helped PSTs attend to the details of practice that other scholars have raised concerns about (Clará et al., 2019). Furthermore, using the protocols that guide lesson study practice across the cycle, but most especially in the post-lesson discussion, helped build shared understanding among the participants about what it means to reflect (Korthagen, 2001).

We have worked together on lesson study efforts in preservice and in-service contexts for 8 years. Across this time, we have developed a deeper understanding of how the *reflect* step pulls together the work that has been conducted by teams in prior steps. Thus, we argue that each step in the cycle feeds into the *reflect* step and that high quality post-lesson discussions and cycle-spanning discussions can feed lesson study cycles whereby PSTs open themselves to the comments of colleagues and generate a greater appetite for tackling difficult problems of teaching and learning.

References

Bryk, A. S., Gomez, L. M., Grunow, A., & LeMahieu, P. G. (2015). *Learning to improve: How America's schools can get better at getting better*. Harvard Education Press.

Chandler-Olcott, K., Dotger, S., Hinchman, K., Waymouth, H., & Newvine, K. (2021). Collaborative design to support digital literacies across the curriculum. In Z. Philippakos, A. Pelligrino, & E. Howell (Eds.), *Design based research in education: Theory and education* (pp. 147–166). Guilford.

Chandler-Olcott, K., Dotger, S., Waymouth, H., Crosby, M., Lahr, M., Hinchman, K., Newvine, K., & Nieroda, J. (2018). Teacher candidates learn to enact curriculum in a partnership-sponsored literacy enrichment program for youth. *New Educator, 14*(3), 192–211.

Clará, M., Mauri, T., Colomina, R., & Onrubia, J. (2019). Supporting collaborative reflection in teacher education: A case study. *European Journal of Teacher Education, 42*(2), 175–191. 10.1080/02619768.2019.1576626.

Dewey, J. (1933). Why have progressive schools?. *Current History*, *38*(4), 441–448.

Korthagen, F. A. J. (2001). *Linking practice and theory: The pedagogy of realistic teacher education*. Lawrence Erlbaum.

Lewis, C., Friedkin, S., Emerson, K., Henn, L., & Goldsmith, L. (2019). How does lesson study work? Toward a theory of lesson study process and impact. In R. Hoang, A. Takahashi, & J. Ponte. *Theory and practice of lesson study in mathematics* (pp. 13–37). Springer.

Penuel, W. R., & Gallagher, D. (2017). *Creating research-practice partnerships in education*. Harvard Education Press.

Lesson Study for Advancing Access and Equity

9

PRESERVICE TEACHER LEARNING ABOUT EQUITABLE PRACTICES THROUGH LESSON STUDY

Melissa Graham and Amy Roth McDuffie

Two common problems in teacher education programs are a disconnect between theory and practice (Putnam & Borko, 2000) and graduates entering the profession ill-equipped to teach diverse populations of students. Lesson study is a promising strategy to address both problems (Parks, 2008). In collaboration with Adele (all names are pseudonyms), the course instructor, we designed and studied an experimental mathematics methods course called the Mathematical Progressions Course (MPC), incorporating lesson study as a key component. We examined PSTs' learning related to equity at each stage of the lesson study cycle. The purpose of the research was to understand how undergraduate PSTs who plan to teach mathematics attended to equity in planning, teaching, and reflecting on their research lesson.

In this chapter, we first discuss literature on the importance of equity in teaching mathematics and our use of lesson study with PSTs. Next, we describe our context and participants and equity-related themes that emerged in our data. We then focus on one group of PSTs and present findings related to their lesson at each step in the lesson study cycle. Finally, we discuss how our findings for the selected group compare to those of other groups and suggest changes for future iterations of courses.

Equitable Opportunities to Learn

Not all students have equitable opportunities to learn mathematics (Copur-Gencturk et al., 2019; Flores, 2007; Nasir, 2016). Factors contributing to inequities include the quality of instruction and the expectations for learning to which students are held. Flores argued that the quality of instruction is a key source of inequities, noting that African American and Latinx students are less likely than white students to access teachers who provide high-quality math instruction. Specifically,

DOI: 10.4324/9781003326434-13

Flores found that schools with high percentages of African American and Latinx students are more likely to have teachers who lack even a minor in their subject areas. Moreover, minoritized students are likely to face low expectations. For example, when teachers estimated students' mathematical abilities, Copur-Gencturk and colleagues (2019) found that non-white teachers favored white students over students of color, and white teachers favored boys over girls, reflecting the persistent problem of race and gender bias. Similarly, Nasir (2016) identified racial stereotyping as a barrier to access and identity development while noting high-quality mathematics instruction could potentially disrupt this issue of unequal access.

Teachers can disrupt inequities by employing strategies focusing on diversity, equity, inclusion, and social/racial justice (DEIJ). Gutiérrez (2007) provided a framework for discussing practices that address equity issues as dominant or critical. Dominant strategies are those in which access and achievement are focused; critical strategies focus more squarely on identity, power, cultural identity, and sociopolitical issues for marginalized groups. Rubel (2017) expanded on the dominant strand by attending to equitable participation and teaching mathematics for understanding. Rubel contended that mathematics instruction should connect to students' experiences. Depending on the context, an activity that connects to students' experiences could be considered dominant or critical. For example, a mathematics activity connected to a shared experience at school, such as passing out valentines or selling items at a bake sale, connects to students' experiences but not necessarily their home or community culture. An activity centered on a family or cultural activity, such as celebrating a quinceñera, that affirms the students' identities, would be considered critical. Rubel expanded on the critical strand by highlighting culturally relevant pedagogy and teaching mathematics for social justice as strategies that support students' identities. Teaching mathematics for social justice examines cultural identity through a lens of power, supporting students in using mathematics to surface and address inequities. Mathematics teachers in Rubel's study who were successful in implementing dominant strategies found critical strategies more difficult to implement.

In working with Adele to design the experimental course, we suggested using lenses specifically created to support PSTs in noticing both dominant and critical equity-based practices (Roth McDuffie et al., 2014). In their study, Roth McDuffie and colleagues (2014) found that these lenses supported PSTs in a mathematics methods course in noticing equitable practices at higher levels (e.g., attending to relationships between teaching moves and student participation with detail and specific evidence). As part of designing the MPC, we suggested these lenses as part of the *study* step of the lesson study cycle because they focus PST noticing on both dominant and critical strategies and provide an opportunity for PSTs to learn about a variety of strategies for attending to equity while teaching. The use of lesson study provided PSTs opportunities to enact the equitable practices they learned about in other activities in the MPC.

Lesson Study with PSTs

The traditional lesson study cycle is often modified to fit a particular context (Lewis et al., 2019). Teachers engaging in lesson study typically *prepare*, *study*, *plan*, *teach* the research lesson, *reflect and revise*, and sometimes *reteach* their lesson (Lewis et al., 2019). Lewis and colleagues included teaching a mock-up lesson as a key component of the *plan* step and noted that this activity most closely approximates actual teaching. Indeed, rehearsals or "mock-up" lessons are common in Japanese lesson studies (Lewis et al., 2019). In adapting lesson study for PSTs in the MPC, we included a rehearsal, which we considered the same as a "mock-up" lesson to provide groups of PSTs with initial feedback on their lesson plan. Our facilitation of rehearsals was like that of Kazemi and colleagues (2009). Adele paused the rehearsal at key moments to allow PSTs to consider their choices on how they might respond to student responses and the potential consequences of those choices.

After the groups were formed and given a topic for their lesson, each group met outside of class to create their initial lesson plan. The research lessons were scheduled during different weeks of the quarter so that each group had an opportunity during class to rehearse their lesson and receive feedback about their lesson and lesson plan (Kazemi et al., 2009). For each rehearsal, Adele assigned PSTs who were not part of the group rehearsing to either take on the role of student or the role of observer. Adele and the PSTs in the observer role paused the group teaching when they identified a teaching dilemma. The feedback and rehearsals were intended to help the PSTs prepare for effective teaching that supported students' learning. Lewis and colleagues (2019) noted that including a mock-up lesson in the *plan* step offered opportunities for noticing that might not be possible without such a close approximation to practice. We found that rehearsals were valuable, but they required time. We typically spent 2 hours rehearsing a 50-minute lesson once.

After working with Adele to design the MPC with these themes in mind, we sought to answer the following research questions:

1 In what ways did PSTs attend to DEIJ in planning their research lesson?
2 In what ways did PSTs attend to DEIJ in enacting their research lesson?
3 In what ways did PSTs attend to DEIJ in reflecting on their research lesson?

Methods

Context and Participants

Adele created multiple opportunities for PSTs to build knowledge (i.e., the *study* step) about DEIJ and equitable practices throughout the MPC. She selected a text, *Taking Action* (Boston et al., 2017), that included equity-based practices in the first chapter that were revisited at the end of each chapter and connected to the content within that chapter. Adele had PSTs view videos, read case-studies, and observe live lessons using equity-based lenses (Roth McDuffie et al., 2014).

At the university where this research occurred, mathematics methods courses in prior years focused on understanding mathematical content, with less attention to pedagogical practice. There were 18 undergraduate students enrolled in the MPC, who intended to teach at the elementary (five PSTs) or the middle/secondary level (13 PSTs). All PSTs consented to participate in this study, which allowed us to collect data from all students. Because we sought rich and nuanced findings, we used case-study methods with purposeful sampling (Yin, 2018).

PSTs in the MPC were at various stages within the teacher education program, with some student teaching in the next quarter and others not yet admitted to the education program. Participants also varied in their grade-level focus. PSTs were in one of the five groups, and each group planned and taught a lesson together, with each group working on a different lesson. We did not specify that all group members would teach part of the lesson, but all groups did so. Groups determined how they broke up the lesson and kept the same roles between the rehearsal and the lesson. Data sources from the PSTs included: Interviews; field notes from class session observations; PST lesson videos; and PST assignments (i.e., lesson plans and reflection papers). We focus our findings on one group of future secondary teachers (grades 6–12) with four members: Cruz, Trent, Sean, and Emma. We selected them because they observed and conducted their lesson study in an Algebra Concepts class that met on campus. To develop our findings, we used data from final interviews and activities related to lesson study.

Data Collection and Analysis

To understand how PSTs attended to DEIJ in the lesson study process, we collected lesson study documents (e.g., PSTs' initial and revised lesson plans, PSTs' post-lesson reflections), researcher field notes during rehearsals and lesson studies, and videos of the research lessons. In the initial analysis, after each rehearsal, we reviewed our field notes and highlighted and identified themes from the discussion during the rehearsal, focusing on themes that attended to equity. For this chapter, we focused on the following two rehearsal discussion themes:

1 Building meaningful, conceptual understanding: Using teaching moves and problems to support students in developing meaningful, conceptual understanding. Developing conceptual understanding is given priority over fluency with algorithms and facts.
2 Supporting equitable participation: Creating opportunities for equitable participation (e.g., opportunities for varied forms of participation and taking up student ideas during group discussions)

These themes are both related to the dominant strands of equity because they center on ensuring students have access to high-quality opportunities to learn mathematics (Rubel, 2017). Specifically, Theme 1 is an example of attending to equity through Standards-based instruction (National Council of Teachers of Mathematics, 2000) and Theme 2 closely relates to complex instruction (Cohen & Lotan, 1995). We focus on these two themes because they were consistent across all groups' lesson study rehearsals. To identify which suggestions from the rehearsal resulted in a change in the enacted lesson, we mapped discussions about teaching dilemmas to changes in lesson plans, as well as to portions of the research lesson video. In video analysis, we created stanzas (Saldaña, 2021), a unit of text that focused on a topic or activity (e.g., lesson launch). Next, for stanzas longer than 5 minutes of recorded video, we broke stanzas down further for more manageable text units for coding. We then compared the lesson video to the rehearsal, and with each opportunity to make a change related to Themes 1 or 2, we recorded whether the PSTs made a change from the rehearsal (and the nature of the change) or kept the lesson as rehearsed. To identify patterns related to DEIJ across the data, we used open and a priori coding (using codes based on the literature described above), and we created analytic memos to describe changes at each stage in the lesson study cycle (Saldaña, 2021).

Findings

As mentioned previously, the research lessons for the five groups were spread out over the 10-week quarter and were on different topics. The group we focus on here was the last group in the MPC to teach their lesson. The students in this group observed and discussed other groups' lessons. They also read and discussed more research and theory on teaching and learning than any other group, in large part because they taught later in the quarter. The MPC discussions and readings likely provided a stronger foundation for their planning and teaching. In other words, this group had more time and experiences than other groups for the *study* and *plan* steps. The group planned and taught a lesson on solving quadratic equations by completing the square. Adele arranged for them to teach their research lesson in the Algebra Concepts course, an introductory algebra course offered at their university.

The Plan Step

The group attended to equity in their lesson plan by including opportunities for multiple forms of participation (e.g., partner work and whole class discussion). In addition, they started their lesson with a roller coaster problem, a context with which they thought the Algebra Concepts students would be familiar. Their initial plan included a first-person video of a nearby rollercoaster that went underground as part of its path. They chose this video for two reasons: (1) It provided an experience to orient students to the context, especially if they were unfamiliar with roller coasters, and (2) the roller coaster passing underground motivated students to find x-intercepts. Sean, who taught the launch portion of the lesson, showed the video and asked students to graph the path of the roller coaster. After the video, Adele paused the rehearsal to express a concern that this context could support a common misunderstanding that a graph is a picture (building conceptual under-standing, Theme 1), and other PSTs in the class suggested alternative contexts. Additionally, some of the PSTs who played the role of students during this re-hearsal had graphs that did not have x-intercepts, and Sean struggled to facilitate a discussion based on their thinking and work that motivated a need for x-intercepts. Again, Adele paused the rehearsal to discuss strategies for broadening participation and listening and responding to student thinking, which was a point of discussion throughout the rehearsal (attending to equitable participation, Theme 2). Table 9.1 indicates which part of the lesson each group member rehearsed and taught, as well as their intended equity focus for that portion of the lesson.

Later in the lesson rehearsal, near the end of class time, Trent and Cruz had not yet practiced their portion of the lesson. Trent indicated that his portion of the lesson did not need rehearsing, "I'm just going to give them problems to work on and go through it" (Trent, Week 9). The group had not turned in a draft of their lesson plan at this point, so Adele asked him to put the problems on the document camera. He showed four examples, and Adele added that four examples were excessive (poten-tially inhibiting students' learning of concepts). Cruz stepped in and discussed his

TABLE 9.1 Overview of the Lesson Components and Focus

PST Leading	Activity	Intended Equity Focus
Sean	Launch the lesson	Draw on students' interests and experiences using a nearby roller coaster to introduce x-intercepts Attend to equitable participation by asking for a variety of student ideas in a large group setting
Emma	Review prior strategies for solving quadratic equations	Build off of students' prior mathematical knowledge to make connections Attend to equitable participation by asking for student ideas in a large group setting
Trent	Pose and facilitate discussion of examples, end with an example that motivates the need for a new technique	Attend to equitable participation by including a variety of examples and asking students to work in pairs
Cruz	Introduce completing the square and connect it to area	Attend to conceptual understanding by connecting the procedure for completing the square to the area of a square

plan to build off Trent's last example, and Adele pointed out that Trent's examples did not produce an intellectual need for completing the square because these examples could be solved by factoring. Adele's questions and comments served to focus the group's attention on students building meaning and conceptual understanding (Theme 1). After the rehearsal, the group revised their examples to create more opportunities for students to share their thinking and to motivate the need to complete the square in solving equations, the mathematical focus of the lesson.

The Teach Step

In teaching their lesson, the example Sean used to launch the lesson was similar to that from the rehearsal (i.e., they kept the roller coaster context, but used a different video). Sean modified the launch to discuss what students noticed and wondered about the roller coaster video to broaden participation (Theme 2). However, even with this change, the PSTs struggled throughout the lesson to encourage participation and facilitate a productive discussion. For example, when Emma asked for student suggestions and probed a student's thinking further, instead of listening and recording the student's ideas, she interrupted and talked over the student during his explanation, limiting students' participation in the discussion.

Emma: So, if we had a times b equals zero (writes $a \cdot b = 0$), what can we say about a and b?

Student: One has to be zero.

Emma: One of them has to be zero? How do you know that?

Student: Because no matter what (Emma began talking, and the student continued his explanation, inaudible).

Emma: (Nods, interrupts, and talks over the student) So, that tells us the zero-product property which says that if a times b equals zero (and then goes on to write the property on the white board).

The group taught a primarily teacher-centered lesson. Moreover, when Trent asked students to work on solving equations by factoring (which had been reduced in number and changed to produce an intellectual need for completing the square), students worked on problems individually at their seats and the room was silent while the group circulated. The group told students that they could work together to solve the problems, but they did not facilitate collaborative work in other ways and did not interact with students. Overall, despite these challenges, we found positive revisions from the rehearsal to the lesson enactment. Specifically, PSTs attempted to probe student thinking on multiple occasions (Theme 2), and the modification of their examples focused more on building meaningful understandings for completing the square than the lesson they planned for the rehearsal (Theme 1). The teacher-centered aspect of this lesson was less prominent during the rehearsal, which we attributed to the way that the PSTs who were playing the role of students engaged in the lesson during the rehearsal. The PSTs asked questions and offered their thinking, more so than the Algebra Concepts students. Indeed, Adele prompted the PSTs to ask questions that they anticipated students might ask. Without the students initiating the interaction, the group teaching the lesson had a more challenging starting point.

The Reflect Step

During the following class meeting, Adele opened the debrief by asking the group of PSTs who taught the lesson to comment first, consistent with lesson study protocols (Lewis et al., 2019). After waiting for the group to comment, another PST in the MPC chimed in, "You took a lot of what students said at face value, without probing" (Jacob, Week 10). PSTs who taught the lesson agreed and added that they were challenged to involve students more due to lack of time. Adele added that it is important in planning to include time to work with students' ideas in order to build conceptual understanding. Adele suggested approaches to build on student ideas more and sequence and connect their examples in a way to improve coherence.

After class, Adele debriefed further with this group of PSTs about their quadratics lesson that focused on completing the square. During this debrief, Cruz referred to the student that "answered all their questions," and the group asked Adele for suggestions on how to facilitate group work so that one student does not take over the discussion (Theme 2). Adele gave several suggestions and then reiterated the strengths of the lesson, including that the "notice and wonder" launch was inviting and the group's effort to plan (e.g., revising the problems) was evident.

Concluding Thoughts

For this study, we examined PST learning about DEIJ and equitable practices as they participated in a mathematics methods course (MPC) incorporating lesson study. The group we focused on had opportunities to learn about equity-directed instructional practices (*prepare/study*) through several activities and readings in the MPC, including activities using equity-based observation tools for noticing (Roth McDuffie et al., 2014) and the course textbook, *Taking Action* (Boston et al., 2017). In the *plan* step of lesson study, the group of PSTs featured here attended to their goal of equity in ways that related to two themes for equitable instruction: (Theme 1) building meaningful, conceptual understanding and (Theme 2) supporting equitable student participation, and these foci were common across other groups of PSTs in the MPC as well. However, planning within these themes differed by groups. Groups who taught their lesson at the beginning of the course had little exposure to readings on equitable practices, so they primarily drew from the textbook in considering how to attend to equity. This challenge of the *study* step may be a potential limitation in involving PSTs that may not be the case for in-service teachers conducting lesson study. The group we focused on in this chapter, who taught their lesson late in the quarter, launched their lesson with a context that they perceived would tie to student interest and experiences to increase student participation and understanding (Themes 1 and 2). This idea, using contexts aligned with experiences (and identities) connected to the equity-based observation tools (Roth McDuffie et al.) and the framing of equity in the text (i.e., draw on multiple resources for knowledge). All groups made substantial changes to their lesson plan following the rehearsal. Common changes involved supporting participation (i.e., writing all student ideas on the board for discussion) and building meaningful understanding (e.g., selecting examples that create an intellectual need for the new mathematical process).

In the *teach* step, all groups made efforts to listen to students as part of attending to participation, and all experienced challenges enacting responsive teaching, potentially due to some PSTs still developing their own understandings of key concepts (Lewis et al., 2019). We saw this when the featured group asked for student input with more open-ended questions, but then they struggled to build off student responses and often took over the conversation. Despite their goals to include students' thinking, the groups experienced first-hand the complexities of enacting these goals in teaching (Resnick, 1987). In their first group meeting, Cruz mentioned that he wanted to anticipate what students would do and "have this memorized." In an interview at the end of the course, when asked what he had learned from teaching a lesson to real people, Cruz laughed and said, "You can't memorize, you're teaching! I tried memorizing what I was going to say. I typed it all up like I'm going to say this, and I pretty much said nothing of what I wrote." In other words, lesson study, and specifically enacting the lesson, exposed the challenging nature of responsive teaching and the need to engage and listen to each student. A central component of equitable teaching is not assuming that one (memorized) way of teaching will meet all students' needs.

The *reflect* step of the lesson study cycle revealed the essential nature of this stage, as well as the importance of the presence of a knowledgeable other (Adele). For the PSTs featured in this chapter, Adele's comments about planning for time to listen to students and for students to contribute and participate were applicable for all groups, consistent with Rubel's (2017) call for increasing equity through multidimensional participation. Before lesson study, PSTs had theoretical conversations about responding to student thinking, but experiencing these lesson studies grounded their conversations in authentic classroom contexts and supported PSTs in connecting theory to their experiences, as explained by Cruz. Enacting equitable practices is a challenge for both practicing teachers and PSTs, and lesson study shows strong promise in supporting teachers at all stages of their careers.

Returning to our original questions regarding ways PSTs attended to DEIJ, we found a limitation that is noteworthy. Consistent with the group discussed here, other groups attended to equity primarily focused on conceptual understanding and participation, dominant strategies for attending to equity that connect to access and achievement (Gutiérrez, 2007; Rubel, 2017). The group we focused on in this chapter also attempted to attend to equity by selecting a context to support student learning, which could be a critical strategy if it affirmed learners' identities. However, the function of the roller coaster example was to orient and engage the students, not necessarily affirm identities. Hence, PSTs in the course made progress in attending to equity while planning, but they did not focus on critical strategies to the same extent. When we discussed this with Adele at the end of the course, she agreed and had come to the same conclusion. She added that in fear of alienating students, we kept equity too comfortable. In future work, we advise mathematics teacher educators to deliberately address issues of justice and identity. A focus on equity might be uncomfortable for teacher educators and for their students, but these ideas need to be surfaced. Additionally, some of the challenges that this group experienced may have stemmed from them starting a lesson from scratch, rather than modifying an existing lesson to attend to students' identities.

References

Boston, M., Dillon, F., Smith, M., & Miller, S. (2017). Taking action: Implementing effective mathematics teaching practices in grades 9-12. National Council of Teachers of Mathematics, Incorporated.

Cohen, E., & Lotan, R. (1995). Producing equal status interaction in the heterogeneous classroom. *American Educational Research Journal, 32*(1), 99–120.

Copur-Gencturk, Y., Cimpian, J. R., Lubienski, S. T., & Thacker, I. (2019). Teachers' bias against the mathematical ability of female, black, and Hispanic students. *Educational Researcher, 49*(1), 30–43.

Flores, A. (2007). Examining disparities in mathematics education: Achievement gap or opportunity gap? *The High School Journal, 91*(1), 29–42.

Gutiérrez, R. (2007). Context matters: Equity, success, and the future of mathematics education. In T. Lamberg & L. R. Wiest (Eds.), *Proceedings of the 29th annual meeting of the North American Chapter of the International Group for the Psychology of Mathematics Education* (pp. 1–18).University of Nevada, Reno, Stateline (Lake Tahoe), NV.

Kazemi, E., Franke, M., & Lampert, M. (2009, July). Developing pedagogies in teacher education to support novice teachers' ability to enact ambitious instruction. In Crossing divides: Proceedings of the 32nd annual conference of the Mathematics Education Research Group of Australasia (Vol. 1, pp. 12–30). MERGA.

Lewis, C., Friedkin, S., Emerson, K., Henn, L., & Goldsmith, L. (2019). How does lesson study work? Toward a theory of lesson study process and impact. In *Theory and practice of lesson study in mathematics* (pp. 13–37). Springer.

Nasir, N. S. (2016). Why should mathematics educators care about race and culture? *Journal of Urban Mathematics Education, 9*(1), 7–18.

National Council of Teachers of Mathematics. (2000). *Principles and standards for school mathematics*. National Council of Teachers of Mathematics.

Parks, A. N. (2008). Messy learning: PSTs' lesson-study conversations about mathematics and students. *Teaching and Teacher Education: An International Journal of Research and Studies, 24*(5), 1200–1216.

Putnam, R. T., & Borko, H. (2000). What do new views of knowledge and thinking have to say about research on teacher learning? *Educational Researcher, 29*(1), 4–15.

Resnick, L. B. (1987). The 1987 presidential address: Learning in school and out. *Educational Researcher, 16*(9), 13–20.

Roth McDuffie, A., Foote, M. Q., Bolson, C., Turner, E. E., Aguirre, J. M., Bartell, T. G., & Land, T. (2014). Using video analysis to support prospective k-8 teachers' noticing of students' multiple mathematical knowledge bases. *Journal of Mathematics Teacher Education, 17*(3), 245–270.

Rubel, L. H. (2017). Equity-directed instructional practices: Beyond the dominant perspective. *Journal of Urban Mathematics Education, 10*(2), 66–105.

Saldaña, J. (2021). *Coding manual for qualitative researchers* (4th ed.). Sage.

Yin, R. K. (2018). *Case study research and applications. Design and methods* (6th ed.). Sage.

10

LESSON STUDY'S CAPACITY TO DEVELOP NOVICE TEACHERS' PRACTICE OF ANTI-RACIST PEDAGOGIES

Curtis A. Taylor, Kristin Komatsubara, and Daisy Sharrock

The teaching profession is constantly evolving, and teacher education programs are driven by the need to extend and renew teacher practice, skills, and beliefs. Many teachers enter the field of education unprepared to effectively teach the curriculum to students of diverse ethnicities and backgrounds (Gay, 2002). This is a glaring equity issue as it is widely predicted by 2050, more than half the US student population will be non-white. Despite the increase of a majority non-white student population, about 80% of the teaching workforce is white and female (Leonardo & Boas, 2021). Such a racial mismatch between teachers and students may yield greater difficulties for struggling marginalized students whose culture and experience differ greatly from their white teachers (Leonardo & Boas, 2021). While students do not need to learn from a teacher with a similar cultural background, teachers need the skills and resources to be culturally responsive educators who use an anti-racist lens in their practice.

Teacher education programs that focus on anti-racist pedagogies allow teachers to develop the necessary tools to create classroom climates that are conducive to learning for ethnically diverse students, participate in cross-cultural communication, analyze instructional materials, and adapt them for adequate representations of cultural diversity and cultural congruities such as developing rich repertoires or multicultural instructional examples (Gay, 2002). Furthermore, teachers who implement anti-racist pedagogies build positive social and cultural identities in their students and demonstrate high expectations of success for their students (Gay, 2002; Ladson-Billings, 1995a, 1995b). In this study, we identified culturally relevant pedagogy (CRP) (Ladson-Billings, 1995a, 1995b, 1997) and frameworks that can be used with a CRP lens such as universal design for learning (UDL) (Fitzgerald, 2020) and teaching for social justice (TSJ) (Esposito & Swain, 2009; Gutstein, 2003) as anti-racist pedagogies of focus. Each of the selected anti-racist

DOI: 10.4324/9781003326434-14

pedagogies values students' cultural and social assets, supports students' academic achievement, and develops students' critical consciousness to analyze and critique the world around them.

Lesson study supports teachers as instructional designers responsive to their students' needs and knowledge and can support the design of lessons that utilize students' cultures and backgrounds (Collet, 2019; Taylor, 2020). It allows teachers to consider the elements of a lesson that effectively contribute to student learning. To prepare novice teachers to serve their diverse student populations, the authors implemented lesson study as a sustained model of inquiry in a year-long teacher education program. This chapter highlights specific structures of lesson study that supported and hindered teachers' self-efficacy and agency in the implementation of anti-racist teaching practices. In addition, we offer possible resolutions for such hindrances.

Context of Study

This study was conducted at the San Diego Teacher Residency (SDTR), a 2-year teacher preparation program housed within High Tech High's Graduate School of Education (GSE). In the first year of enrollment, SDTR residents complete course-work with a cohort, and spend 4 days a week as a student teacher in one of High Tech High's K-12 schools to earn their Multiple Subject or Single Subject California Preliminary Teaching Credential. The second year of SDTR supports candidates as they move into their teaching careers. While working as fully employed teachers, residents return to High Tech High's GSE once a week to complete their M.Ed in Teaching and Learning. They build on the learning and the relationships formed during their first year of residency and explore shared problems of practice that emerge in their classrooms. The program is designed to support its teachers in designing and teaching through projects grounded in anti-racist pedagogy.

Participants of this study were 21 novice teachers enrolled in the second year of the M.Ed Teaching and Learning program at SDTR. This 1-year program engaged teachers in three lesson study cycles. Each 12-week cycle, teachers developed a research question and lesson hypothesis related to teaching and learning in their context, analyzed and synthesized relevant research and literature related to their research question and lesson hypothesis, and designed and implemented a research lesson followed by a debrief and consolidation of learning. Additionally, each cycle focused on a different anti-racist theoretical framework. The first cycle focused on the theoretical framework of CRP, the second cycle focused on UDL with an anti-racist lens, and the third cycle focused on TSJ. The authors hypothesized that focusing each lesson study cycle on an anti-racist pedagogy, teachers would develop the dispositions and practices of an anti-racist educator. We started with CRP to establish an understanding of student-centered practices and concluded our year on TSJ to expand on the principles of CRP to build capacity for connecting lessons to community and the broader world.

Teacher Agency

Anti-racist pedagogies represent a set of skills that, in our opinion, require teachers to be willing to learn, try, fail, and try again. The degree to which a teacher is willing to engage in this process is supported by their sense of agency. Agency is broadly defined as acting to produce a particular outcome and is supported by a teacher's belief that they can be successful, and that the task or outcome is relevant to their goals. In this study, we conceptualized "agency" as a construct where novice teachers felt confident they could meet the learning needs of their students, having a sense of belonging to a learning community, and that they felt what they were learning was relevant to their current and future goals (Ryan & Deci, 2000). We further conceptualized agency as socio-constructivist in nature. A novice teacher's sense of agency is influenced by past experiences, their hopes for the future, and most importantly, the daily interactions with their environment—their colleagues and students (Emirbayer & Mische, 1998).

Research Questions

The research questions that guided this study were: How does lesson study as an inquiry model support or hinder…

- Novice teachers' implementation of anti-racist pedagogies such as CRP, UDL, and TSJ into practice?
- Novice teacher agency in the implementation of these anti-racist pedagogies?

Methods

Teachers in the M.Ed program at SDTR participated in three lesson study cycles that engaged them in learning and experiencing CRP, UDL, and TSJ. All M.Ed teachers were invited to participate in a 15-minute anonymous agency survey at the beginning of the year. In the last class of each trimester, consenting teachers completed the same teacher agency survey. Additionally, five teachers were invited to participate in a semi-structured interview at the end of each trimester, which corresponded with the conclusion of a lesson study cycle. Teacher interview data and agency survey scores were collected and analyzed to determine the factors, experiences, and complexities that may shape a teacher's sense of agency in implementing anti-racist pedagogies. Descriptive statistics were used to analyze the teacher agency surveys and teacher interview data was transcribed and analyzed with thematic coding. As teachers described their experiences from lesson study, particular attention was given to their references and capacity related to CRP, UDL, and TSJ.

Findings

Findings related to the first research question reveal how lesson study in a teacher preparation program supported and hindered novice teachers' implementation of anti-racist pedagogies. One support of lesson study was that it provided teachers with a safe collaboration space. In addition, the lesson study's *study* step supported teachers' development as teacher researchers and thus strengthened their understanding of anti-racist pedagogies. In contrast, a unique finding of this study is how the lesson study model hindered teacher implementation of anti-racist pedagogies. Teachers reported two hindrances related to the lesson study model. Learning the lesson study model was challenging at first for all the teachers in the study. In addition, some teachers struggled to navigate the delicate process of establishing trust as a lesson study team. Current research shares much about the support and successes of lesson study in teacher preparation programs. While our research has similar findings on lesson study success, our work emphasizes the hindrances our teachers experienced and possible resolutions for each. All names of participants are recorded as pseudonyms.

Lesson Study Supports

A Safe Space to Collaborate

Our interest in how lesson study supported teacher implementation of the anti-racist pedagogies CRP, UDL, and TSJ began with an analysis of teacher interview data. We noticed a pattern in the data, through open and thematic coding, that illuminated how much teachers valued the collaborative space of lesson study. Because of the collaborative nature, lesson study provided teachers with a sense of safety in exploring the new pedagogies they were learning, such as trying out new practices, receiving peer feedback, and learning from different perspectives. A few conditions were put into place to allow teachers to feel safe in their lesson study teams: (1) All teams co-created norms on expected ways of being and doing together, (2) every team started every meeting with an icebreaker to reflect on their own personal/professional celebrations and grapples, and (3) teams met on a weekly basis that afforded more connection to each other. Conrad, a teacher enrolled in the program, shared:

> I think one of my happiest places is sitting at a round table in person, or virtually, and just like ideating and throwing out ideas and talking through things with my peers … the process has been really valuable in terms of helping me feel safe to try things and to know that I have a support team and resources that I can go to when I feel confused. So, that helps with my confidence a lot.

A safe space is necessary when building teacher capacity in anti-racist pedagogies. Because of the collaborative nature of lesson study, Conrad experienced an increase in his confidence in trying out new pedagogical practices. In addition, he felt that he had the support system of his colleagues who provided him clarification when needed and resources. Riley corroborated Conrad's feelings as she appreciated having, "the space to collaborate with someone else, with different backgrounds, is just the wealth of knowledge that is needed." Becoming an anti-racist educator requires teachers to listen, understand, and experience various perspectives. This allows teachers to translate a sense of belonging and inclusivity into their classrooms.

Another imperative aspect of building capacity in anti-racist pedagogies is the act of vulnerability. The power of a safe space allows teachers to be vulnerable with each other and allows teachers to reflect on areas of growth and need. Lesson study afforded teachers a space to be vulnerable as they learned more about their identity as educators, tested and tried out new practices during the *study* step, and opened up their classrooms for observations during the *teach* step. Antonio shared:

> We have that comfort level to push each other into places that maybe we didn't think about before…I can be vulnerable [with them]. And, I think that's what's great. Because, I wouldn't be able to really express how I feel about things or push myself [to say], "Hey, you might have missed this in the lesson."

Antonio's recognition of his own vulnerability, and that of his peers, afforded him the space to push on the team's practice. Additionally, teachers experienced a sense of belonging as they listened to one another during their lesson study cycles. Wallace, a teacher enrolled in the program, stated, "I think that when you do lesson study, it feels more personal. It feels like you're contributing. It feels like you're hearing about other people's experiences and why those contributions work." Not only did lesson study provide the teachers with a safe space, but it allowed them to build their capacity as researchers.

Lesson Study Develops Teacher Researchers

The *study* step of lesson study developed teachers' awareness of the research on CRP, UDL, and TSJ, deepened their understanding of students, and established their ability to put research into practice. Teachers often lack the time to read research books and journals or may be unable to make the research practical. This is even more pronounced for new teachers entering the field, but lesson study provided the necessary scaffolds to support teachers in developing these skills.

Research is a core component of lesson study. Lesson study's *study* step contributed to the teacher's awareness of the available research on anti-racist pedagogies including CRP, UDL, and TSJ. Teachers reported their appreciation for the process of finding research to support their understanding of a new practice. When asked what supported her implementation of culturally responsive pedagogy, Jennifer shared, "I found myself really excited about the research. Being able to find a reason that supported something that we wanted to do or [we] had this aha moment and say, 'Ok, this is why we should do this.'" Jennifer's colleague, Kate, also highlighted how the *study* step was impactful to her practice of CRP, "Having to do a research-based study has been helpful because I don't think that's something that we [would] necessarily do on our own … like do research and then directly apply a specific [CRP] practice in our classrooms."

Furthermore, teachers could take theoretical research and make it practical. During the *study* step, each lesson study team identified a research question of focus (e.g. *How do we support students in trusting their mathematical thinking?*). This guided them in finding research that identified CRP, UDL, or TSJ practices to try in their research lessons. When asked the question, "What about the lesson study model did you find the most helpful in learning about CRP, UDL, or TSJ?" Marcus, appreciated this step of lesson study by sharing:

> I think the research [on TSJ] is really fascinating. I like how the [lesson study] model is set up where you can pose the research question, and then find the current literature or what the peer-reviewed studies have to say. Then, being able to use those practices in your classroom.

Thomas shared his success with practices he identified through the research in one of the lesson study cycles, "Something that came up in my research was [mathematical] revisions. It's a big part of my class now." He also shared identifying mathematical norms to support his students in group work, "I found these four or five math norms and how to have productive classroom discussions." Although teachers felt supported through the *study* step, they also experienced some hindrances with the lesson study model.

Hindrances with Lesson Study

Learning the Lesson Study Model Was a Challenge

Learning to teach through an anti-racist lens can be a major shift in practice for new teachers. In addition, learning a new inquiry model like lesson study can add extra risk to the already complex art of teaching. Dealing with multiple novel factors, teachers expressed that they found it challenging to comprehend the lesson study process in tandem with learning a new pedagogy. For example, Jennifer expressed this challenge by sharing, "I think that a little bit is lost with the content because you're spending so much time trying to figure out lesson study." Learning the process of lesson study not only impacted the teachers' depth of understanding, but also their sense of confidence in the implementation of the new pedagogies. Mario and his lesson study team experienced this challenge while exploring complex instruction. Complex instruction is an instructional approach where teachers use cooperative learning to teach at high academic levels in a diverse classroom and is included in the nested framework for Teaching Mathematics for Social Justice (Berry et al., 2020; Cohen et al., 1999). He shared, "We spent so many weeks focused on making the lesson, drafting our lesson hypothesis, and doing much more that we never really understood complex instruction."

Establishing Trust as a Lesson Study Team

Another major hindrance our teachers faced was establishing trust in their lesson study teams. Hoy and Tschannen-Moran (1999) define trust as reducing uncertainty and having confidence that our expectations of others will be met. Trust is the foundation for cohesive and productive relationships in professional learning communities. Unfortunately, not all the teachers in this study felt a high sense of confidence that expectations would be met in their lesson study teams. For example, Jennifer expressed her uncertainty about roles and responsibilities in her lesson study team:

> It comes down to the fact that certain things have to be created for the lesson. That, to me, feels like unclear territory. We are teachers and we are all creating material for our own classrooms, so should the host teacher be the person creating the material [for the research lesson], or are we all designing it together, or do we all take on a piece of it? I don't know the answer to that.

Jennifer's uncertainty around the roles and responsibilities of the team, and conflicting priorities in their respective classrooms, impacted her commitment to the team's ability to attend to the logistical components of implementing a research lesson. Moreover, some teams experienced challenges when it came to communication. This was illuminated as Conrad shared, "I think consistent communication was difficult because of what we all had going on separately." A lack of effective communication on a team can negatively impact the amount of trust in a team. Thus, the level of trust on a team can shift a team's focus to internal group dynamics, and negatively impact how effective the teachers are in implementing anti-racist teaching practices.

Possible Resolutions for Hindrances

With any new learning model, there is an expectation that challenges will arise. Through feedback, reflection, and additional research, the authors have found possible resolutions to the hindrances mentioned above. Possible resolutions include providing teachers space and time to understand the lesson study model and creating a culture of generative conflict.

First, teachers need time and space to understand the lesson study model. Lesson study is a simple idea (improve instruction through collaboration, plan a research lesson, and record the impact on students) but a complex process (Lewis & Hurd, 2011) that is unfamiliar for many teachers, including novice teachers. A possible resolution to this issue is to spend a trimester, or a minimum of 8 weeks, on facilitating the process of lesson study with new teachers. We believe the best method to understand and build capacity for lesson study is for teachers to experience the process. This allows teachers to understand each step of lesson study deeply. Moreover, it provides teachers space and time to deeply understand the process by analyzing prior lessons study examples, drafting a research question that is reflective of a common problem of practice, identifying focal students, studying content and curriculum material (also known as *kyouzai kenkyuu),* drafting a lesson hypothesis with identified teaching practices, and collecting student data to inform next steps in instruction.

Through lesson study, we aim to build a culture where generative conflict is a natural work process. Conflict is a natural consequence of team collaboration, but how we navigate conflict can benefit the quality of work that is produced and can deepen the level of trust in a team. Conversely, it can interrupt the learning process, diminish the quality of work produced, and fracture trust in a team. In creating anti-racist and liberatory learning cultures, we must be open to generative conflict. Generative conflict requires the courage and know-how to engage in conflict to name and disrupt harmful dynamics and make space for healthier dynamics to emerge to create stronger relationships, greater trust, and mutual growth and learning. For example, lesson study team members can participate in dialogic interviews, an informal conversation between partners where each partner shares personal stories, beliefs, and experiences. Dialogic interviews open up the possibility of having richer, more informative interactions where all teachers are validated and understood (Collins, 2016). By participating in dialogic interviews, teachers begin to develop vulnerability with each other which leads to a sense of trust.

Teacher Agency in Implementing Anti-racist Pedagogies

Findings related to the second research question were derived from the pre- and post-intervention agency survey, interview data, and documentation analysis. Survey results collected at the end of each trimester suggest that teachers found the SDTR lesson study model relevant for achieving their goals of supporting the learning needs of their students and that they more regularly used student-generated data to inform their teaching practice. Interview data corroborated these findings.

Pre-and Post-intervention Agency Survey Analysis

The percent of agree and strongly agree survey responses were compared between the pre-and post-intervention surveys. A one-tailed homoscedastic t-test analysis was used to determine the level of significance for shifts in the teachers' sense of agency. Eighteen teachers took the pre-intervention survey and 17 of the 18 completed the final post-intervention survey.

For nine of the ten agency survey questions, the percentage of teachers answering "agree" or "strongly agree" increased between the pre- and post-intervention surveys. For the question "My SDTR cohort feels like a community that is always trying to get better at meeting the needs of our students," the pre- and post-percentages of teachers responding "agree" or "strongly agree" were 72% and 71%, respectively. Given that the teachers in the cohort were all novice teachers, it is plausible that even in the early days of the program they felt that members of their cohort were focused on getting better at meeting the needs of their students. This may also have been reinforced early on through the strongly student-centered program focus.

While none of the questions reached a *p-value* indicating 95% confidence that the null hypothesis is rejected ($p \leq .05$), several questions were in the 90% confidence range, including: "I use data such as student work, exit cards, or student surveys to determine if my teaching practices are working for all my students" ($p = .099$), "The professional development I receive through the SDTR lesson study model helps me meet the learning needs of my students" ($p = .096$), and "The professional development I receive through the SDTR lesson study model helps me improve my teaching practice" ($p = .098$). These results suggest that teachers were more likely to use student-generated data in the form of surveys, student work, or exit tickets to evaluate their teaching practices after completing the SDTR program and that they believed the SDTR lesson study model helped them meet the learning needs of their students and improve their teaching practice.

Concluding Thoughts

Schools are microcosms of our larger society that serve as sites promoting the interests of the dominant culture (Esposito & Swain, 2009). For our schools to be welcoming, safe, and anti-racist spaces for all children, teachers must shift their pedagogy to create inclusive environments that value the assets of our low-income, culturally, and linguistically diverse students. With this transition, teachers may experience challenges resulting in a loss of teacher self-efficacy and agency. Yet teacher efficacy and agency are critical to reform and have been highlighted by research to have positive outcomes related to teacher instruction, confidence, and motivation (Ryan & Deci, 2000).

In sum, this study supports earlier positive findings on lesson study with novice teachers while offering new insights into the hindrances of the approach, particularly around teachers' sense of agency to implement anti-racist pedagogies. This study found that lesson study influenced teachers' implementation in two main ways: (1) Teachers reported the potential of lesson study providing a safe space for collaboration, and (2) it supported teachers as teacher researchers during the *study* step. We recommend further research on how trust develops within lesson study teams and how that could impact teachers' agency to implement a new pedagogy into their practice. Last, teachers valued the SDTR lesson study model and found it relevant to their goal of supporting student learning. For teachers in the M.Ed program, the lesson study approach likely reinforced an established and shared aim of student-centered instruction.

Anti-racist teaching is more important now than ever. COVID-19, as well as the nation's political landscape, has revealed and exacerbated injustices already present in K-12 education. While this can paint a bleak picture of classrooms, interviews with focus teachers demonstrated the resilience, motivation, and commitment of novice teachers to learn and implement anti-racist pedagogies. Furthermore, the lesson study approach demonstrates the collaborative potential of teachers to create collective communities committed to liberatory practices.

References

Berry, R. Q., Conway, B. M., Lawler, B. R., & Staley, J. W. (2020). *High school mathematics lessons to explore, understand, and respond to social injustice*. Corwin Press, Inc.

Cohen, E. G., Lotan, R. A., Scarloss, B. A., & Arellano, A. R. (1999). Complex instruction: Equity in cooperative learning classrooms. *Theory into Practice, 38*(2), 80–86.

Collet, V. S. (2019). *Collaborative lesson study: ReVisioning teacher professional development*. Teachers College Press.

Collins, M., "The Dialogic Interview: Co-Creating Meaning as an Alternative to Discovering 'What Is'" (2016). Undergraduate Research Conference (URC) Student Presentations. 24. https://scholars.unh.edu/urc/24

Emirbayer, M., & Mische, A. (1998). What is agency? *The American Journal of Sociology, 103*(4), 962–1023.

Esposito, J., & Swain, A. N. (2009). Pathways to social justice: Urban Teachers' uses of culturally relevant pedagogy as a conduit for teaching for social justice. *Penn GSE Perspectives on Urban Education, 6*(1), 38–48.

Fitzgerald, A. (2020). *Antiracism and universal design for learning: Building expressways to success*. CAST Professional Publishing.

Gay, G. (2002). Preparing for culturally responsive teaching. *Journal of Teacher Education, 53*(2), 106–116.

Gutstein, E. (2003). Teaching and learning mathematics for social justice in an urban, Latino school. *Journal for Research in Mathematics Education, 34*(1), 37–73.

Hoy, W. K., & Tschannen-Moran, M. (1999). Five faces of trust: An empirical confirmation in urban elementary schools. *Journal of School Leadership, 9*(3), 184–208.

Ladson-Billings, G. (1995a). But that's just good teaching! The case for culturally relevant pedagogy. *Theory into Practice, 34*(3), 159–165.

Ladson-Billings, G. (1995b). Toward a theory of culturally relevant pedagogy. *American Educational Research Journal, 32*(3), 465–491.

Ladson-Billings, G. (1997). It doesn't add up: African American students' mathematics achievement. *Journal for Research in Mathematics Education, 28*(6), 697–708.

Leonardo, Z., & Boas, E. (2021). Other kids' teachers: What children of color learn from white women and what it says about race, whiteness, and gender. In M. Lynn & A. D. Dixson (Eds.) *Handbook of critical race theory in education* (pp. 313–324). Routledge.

Lewis, C. C., & Hurd, J. (2011). *Lesson study step by step: How teacher learning communities improve instruction*. Heinemann.

Ryan, R. M., & Deci, E. L. (2000). Self-determination theory and the facilitation of intrinsic motivation, social development, and well-being. *American Psychologist, 55*(1), 68–78.

Taylor, C. A. (2020). *Bringing the culture to mathematics: The impact of lesson studies on math Teachers' understanding and self-efficacy of culturally relevant pedagogy*. University of California.

11

PROMOTING TORRES' RIGHTS OF THE LEARNER WITH ELEMENTARY PRESERVICE TEACHERS THROUGH LESSON STUDY

Crystal Kalinec-Craig, Keely Hulme, Karisma Morton, Colleen Eddy, Fardowsa Mahdi, Dittika Gupta, and Mark S. Montgomery

The mathematics education community has persistently called for promoting equitable mathematical learning experiences for children in grades PK-12. According to the National Council of Teachers of Mathematics (2014), "all students [should] have access to a high-quality mathematics curriculum, effective teaching and learning, high expectations, and the support and resources needed to maximize their learning potential" (p. 59). Calls such as these are indicative of the increased recognition that many students, particularly girls, BIPOC (Black, Indigenous, People of Color), and those who face socioeconomic challenges, continue to be marginalized in mathematics classrooms (Goffney et al., 2018; Joseph et al., 2017). This marginalization often begins early in students' mathematics learning experiences whereby mathematics learning advantages some children (namely white and affluent) while BIPOC students continue to be pushed farther to the margins (Kerckhoff & Glennie, 1999). Consequently, practitioners, researchers, policymakers, and other stakeholders have explored ways to address this issue.

Mathematics methods courses are key locations where equity in mathematics learning and teaching can be addressed directly. Within these courses, mathematics teacher educators (MTEs) can provide rich and meaningful learning experiences that support preservice teachers (PSTs) as they transition from learners of mathematics to teachers of mathematics. Those meaningful opportunities include moments for PSTs to reflect on their mathematics learning and explore if (and/or how) that impacts their emerging practice. Use of lesson study during mathematics methods courses helps educators reflect, collaborate, and improve instruction through a cycle of *studying, planning, teaching*, and *reflecting* (Lewis et al., 2019). Lesson study also allows MTEs to improve instruction for the PSTs (students) with a specific goal in mind. Most elementary PSTs are white women (Schulte, 2009) who need to experience more rich and meaningful learning opportunities as PSTs

DOI: 10.4324/9781003326434-15

to create similar learning opportunities for their own students who bring a diverse set of cultural and linguistic assets to the classroom. By diversifying the profession in terms of teachers' experiences and identities, the work of teaching mathematics is more aligned with how children from all cultural and linguistic backgrounds learn and use mathematics in their homes and communities (Hollins, 2011). It is therefore important for mathematics methods courses to support all PSTs as they learn to identify those practices that resonate with their professional stance while learning to incorporate more practices that are equitable, anti-bias, and embrace the humanity of the children in their classrooms.

As such, our chapter focuses on how MTEs used lesson study to collaboratively create and improve a lesson to engage with PSTs in equitable mathematics instruction utilizing Torres' Rights of the Learner (RotL) framework (Torres, 2020). The research team consisted of seven MTEs at four universities in the south-central region of the US, two of them being doctoral students with a range of years of experience teaching elementary math methods. As this study took place during remote instruction due to the COVID-19 pandemic, we provided much support to each other as fellow MTEs and established ourselves as virtual collaborators (Gupta et al., 2022). The goal was to build learning experiences that model what equitable mathematics instruction can look like to disrupt the pattern of inequitable and dehumanizing teaching practices in elementary classrooms. The rest of this chapter describes our lesson study process which includes how we engaged in the four steps of *study*, *plan*, *teach*, and *reflect* (Lewis et al., 2019). The chapter concludes with implications for future research in mathematics teacher education.

Step One: Study

We began our lesson study by discussing the integration of equity in each of their methods classes as a means to increase our collective knowledge base. After some discussion, we decided on Torres' (2020) RotL framework to engage elementary PSTs enrolled in a math methods course in a group-worthy problem-solving task (Cohen & Lotan, 2014) involving finding the area of nonstandard objects using nonstandard measurement. As a group, we decided on the equity-focused theoretical framework, pedagogical mathematics tool, and mathematics content area that formed the foundations of the research lesson.

Torres' RotL Framework and Group-worthy Tasks

Torres (2020), the bilingual elementary teacher who theorized the RotL, argues that teachers (including MTEs) should always ask students "What are the learning conditions that will help you to be the best learner that you can be?" and serves as one way to introduce the RotL to students as they learn to recenter their learning to their needs. When there is a foundation of trust and mutual respect among students and their teachers, there are more meaningful and sustainable learning opportunities in mathematics classrooms. As such, Torres' RotL framework is based on students' civil rights as learners of mathematics, which includes the right to be: (1) Confused; (2) claim a mistake and revise your thinking; (3) speak, listen, and be heard; and (4) write, do, and represent what makes sense. When teachers emphasize the first RotL in their practice, they acknowledge the ways that children learn complex skills and concepts; learning is messy and is not always straightforward. When children express their confusion, they acknowledge a moment in their thinking that requires resolution to move to a conclusion or an answer. The second RotL is typically one that traditional mathematics classrooms have demonized as something to avoid, claiming a mistake. When children exercise this right, they acknowledge their mistakes will not help them answer the question and need to contemplate a new direction to solve the problem. The first and second RotL encourage perseverance and productive struggle in the ways that humans engage in solving a problem using mathematics and their own life experiences.

The third and fourth rights are intertwined, and the reader might notice the parallel structure of the rights (Torres, 2020). The right to speak, listen, and be heard uses the auditory and vocal senses we use when communicating our thinking with another. When children learn that the voices of their classmates are just as important as their own, they develop active listening skills beyond just waiting to speak next. Finally, the fourth RotL is one in which children learn other means of communication (e.g., writing, drawing, and gesturing) are just as valuable as listening and speaking with each other.

Each of these rights is important for all students. Still, they are especially valuable for children who are recent immigrants and are unfamiliar with the traditional algorithms promoted in the United States, for children learning mathematics in multiple languages, and for children who have not experienced a classroom rooted on trust for and with each other (Kalinec-Craig, 2014).

One practice that supports the RotL is the notion of group-worthy problem-solving tasks. As described in the research of complex instruction (Cohen & Lotan, 2014), group-worthy problem-solving tasks have key elements such as being open-ended, having unclear answer or solution strategies, and allowing the use of multiple strategies to approach the task collaboratively. The research on group-worthy problem-solving tasks has shown that the structure encourages students to collaborate equitably, engage in active listening, and acknowledge the strengths of each other's thinking when solving a complex task. Group-worthy problem-solving tasks, with appropriate entry points for students to begin the task, imply that each student has something to contribute to the task and solution; the RotL is one way to encourage this productive group work.

Step Two: Plan

The MTEs prepared a lesson by planning the application of the RotL through a group-worthy problem-solving task involving filling a region with five different non-standard shapes (e.g., teddy bears, hearts, ovals, and stars). The task required

PSTs to determine and explain how much area was covered by the shapes. The purpose of the task was to encourage PSTs to utilize the RotL among themselves as they completed the task and to consider which shape might fill more of the region (as compared to using a standardized unit of measurement such as a square centimeter or inch). Since the task involved non-standard shapes, each group had to collaboratively find a strategy to solve this open-ended problem with an unclear answer or solution strategy.

We were specifically interested in addressing the following research question: In what ways did our lesson study about a group-worthy, problem-solving task support PSTs' awareness and use of Torres' RotL? PSTs consented through IRB protocols and all names are pseudonyms to protect their identity.

Step Three: Teach

The unit was developed and refined using the lesson study process (Lewis, 2002). The initial lesson was developed, observed, and studied by the MTEs, and revised following each of the three iterations of the lesson's implementation based on observation and discussion of each iteration. Each iteration of the lesson was implemented in three mathematics methods courses with PSTs by two different MTEs at one of the four universities. Table 11.1 shows the progressive iterations of the lesson study and the revisions.

Three main sources of data were reviewed from the implementation of the research lesson. First, PSTs read a description of the RotL framework and its implementation in an elementary mathematics classroom (Kalinec-Craig &

TABLE 11.1 Lesson Study Iteration

Iteration 1	Iteration 2	Iteration 3
Matching review activity of RotL	Review and show the RotL	Review of RotL
Three videos of student experiencing RotL	Two videos of student exercising RotL	Two videos of student exercising RotL
Third video Jamboard in groups		
PSTs engage in non-standard measurement activity	PSTs engage in non-standard measurement activity	PSTs engage in non-standard measurement activity
Conclusion and discussion of activity around Smith and Stein's 5 Practices of Mathematical Discussion	Discussion of activity • Debrief 1—reviewed other strategies • Make changes • Debrief 2—strategies, revisions, and RotL	Discussion of Activity • Debrief 1—convince why estimation was closest (strategies) • Make changes • Debrief 2—revisions and RotL exercised

Robles, 2020). They responded to the reading in an online discussion by indicating what aspects resonated with them and by asking questions of their fellow PSTs. The second and third sources were from transcripts of the synchronous whole class session where PSTs watched Cognitively Guided Instruction videos (Carpenter et al., 1999) of a third grader engaged in mathematical problem-solving tasks and identified and discussed what RotL he exercised while solving the problems. Then PSTs were divided into groups to solve a measurement task that required them to determine the area of a shaded region using non-standard figures with the known area. After completing the task, they were then asked to reflect on what RotL they exercised as they were completing the task. For final reflections, PSTs stated what they learned from the whole class synchronous session.

The major revisions because of reflection based on observation and debrief were threefold:

1 We chose to only show two videos and not three of children solving mathematics problems and engaging in the RotL to allow more time for the mathematical task.
2 The discussion focused on PSTs' thinking about the mathematics as it related more to RotL than the discussion on 5 Practices of Mathematical Discussion (Smith & Stein, 2018).
3 The whole group discussion emphasized more specific questions that encouraged PSTs to convince the MTE as to why their answer was the most accurate for describing the region using the non-standard units of measure.

With each iteration of the lesson, the MTEs followed the lesson study protocol by focusing on equitable teaching practices using RotL and finding area using non-standard measurement to better understand how the PSTs awareness and use of the RotL could improve.

Discussion and Findings

Analysis of the three iterations of the lesson study showed how PSTs discussed the RotL, offered multiple solution strategies, and (re)considered revisions to their thinking based on questions asked by the MTEs.

PSTs Discuss the RotL

At the start of each iteration of the lesson, the PSTs watched a video Austin solving three problems (Carpenter et al., 1999) (one multiplication problem and two division problems). This video was chosen because it demonstrated three of the four rights of the RotL. Austin solved the problems by using his fingers (RotL 4) and was able to explain his thinking (RotL 3). He also recorrected his thinking when he noticed he had made a mistake in his counting (RotL 2). After watching the video, the PSTs discussed which rights they noticed Austin using as shown in the PST statements below.

Moana: It's the right to not immediately comprehend everything we hear and see. [First iteration].

Lucy: Yeah, he's just working it out in the way that he could. And it just took him a bit longer but that's all right. [Third iteration].

Zoe: Yes, so when he went back to count the ladybugs and he did say at first that he was confused. And then he said he counted them and then he was like, oh wait, I made a mistake, and then that's when he went back. Oh, that's when the lady asked him to recount them. [Third iteration].

In seeing the video of Austin using his fingers in the context of the RotL 4, they noted the power of using their fingers as a means of supporting children to engage in mathematics in ways that make sense to them:

Dia: I think it would be okay (to use their fingers to count) because you know the right, it says "to write, do, and represent what makes sense to you." And that's what makes sense to him. So, I think taking that away from him would be a detriment to him. [Second iteration].

The RotL were no longer something that PSTs were reading in an article or posted on a wall in a classroom. By actively observing a student exercise the RotL with a mathematics problem, PSTs discussed RotL, what the rights look like with an elementary student and considered the consequence of not having the RotL when doing mathematics.

Purposeful Questions from MTEs Encouraged PSTs to Engage in the RotL

After watching the videos of Austin exercising his RotL, PSTs worked in small groups on the group-worthy problem-solving task regarding measuring a region with non-standard units of measure. In the first iteration of the lesson, the MTE asked broad questions about what the PSTs noticed when solving the problem and what they noticed when examining the work of the other groups:

MTE 1:   I want you right now to go take a walk, go check out the other gems and see what the other groups are doing. So, give you 3 minutes to just window shop. Make note of what the strategies appear to be, how they're similar different to what your group strategies are. [First iteration].

One example of the PSTs' responses from the first iteration was:

Blossom: Okay, so our strategy was that we just put the oval in the purple part [of the given figure] until they all fit in there. And then, we multiplied them by their area and that got us the equation over to the left. Our strategy did not change once we looked at others [shapes]. The right we exercised was one. We were confused when we started.

The debrief of the lesson began with the MTE who taught the lesson followed by the observers. The MTEs noticed that they needed to revise their questions to further elicit more information from the PSTs as to their thinking and ways of solving the task:

And then I didn't feel as though the questions that were … so what ended up happening was the students ended up explaining their process for solving the problem. Which was fine. I just felt in the moment that there was more that I could have done or more that the lesson could have done so kind of wrap that up a little bit here.

MTE 1, debrief from first iteration

As a result, the MTEs asked the PSTs in future iterations to focus on justification to support their rationale for why their answer was the most accurate. For example, the MTE asked the PSTs to not only share what their strategy was but also, "How can you convince me that your answer is the best fit?" Upon the first whole group share out, one PST shared their strategy for justifying their solution and exercised their third and fourth RotL:

Mallory: We decided to use the ovals because we thought that they take up the most space and they're also more convex, as opposed to concave so it covers more or less space without it being dead area. So we did four across and we did six down. So, and then we subtracted to ovals from the center. So we did four times six is 24 subtract two is 22 times seven, and gives us 150 [sic] for the region.

To encourage the PSTs to consider the strategies of other groups and exercise their second RotL, the MTE encouraged the PSTs to make any necessary revisions based on whole group discussion, why other groups chose that strategy, and which RotL they exercised and an example of how they exercised that right.

After the PSTs concluded their second return to the task to review and revise their thinking, there were some PSTs who revised their solutions.

Catherine: Okay, so yes, we revised ours by still doing the bears. But we added half bears to fill in more space. So, we just counted that as eight

Square unit. Or eight units and then it gave us a bigger square units number. So, we kind of identified the mistake and then used right number four, which I believe it's like you can identify and explain it. [Third iteration].

For Catherine's group, they exercised not only the third and fourth RotL, but also explicitly exercised the second RotL by using only half of the shape (the teddy bear), which is still a shape in itself. On the other hand, other PSTs did not revise their thinking when returning to their breakout rooms and chose to stand with their original strategy; in this case, they exercised their third RotL.

Zoe: And so, we didn't really change anything in the square. Just because we thought that the hearts did a pretty good job at covering most of the space. We didn't think any other shape would be able to do that. We exercised all for them [the RotL]. But I would say the first one. So, the right to be confused. And then the fourth one to do something that would make sense to us. [Third iteration].

Ultimately, the PSTs offered more concrete explanations compared to their original solution strategies. Also, because the MTE explicitly encouraged the PSTs to return to their small groups and consider revisions to their strategies, this offered an opportunity for some groups to exercise their second, third, and fourth RotL.

PSTs Engaged in Self-Reflection through Groupwork

At every iteration of the lesson study, the PSTs considered how and why their shape and orientation would best fit the purple region. Once the PSTs considered the initial selection of a shape, they also considered if there was another shape that might offer a better fit to fill the purple region. Although some shapes filled in the purple region better, some groups noticed their strategy might have made it more difficult to calculate the accuracy of their estimation. In observational notes from one MTE, the group used math terms like convex versus concave to describe and justify their use of the ovals.

"Do we need the area of the white part?" They then started exploring white space area. They changed the orientation of the ovals from horizontal to vertical and came up with 14 sq. units but never did anything with that new information … When talking about the rights of the learner, they decided that they had used all of them but they chose to highlight the right to make a mistake (orientations of ovals in the white space). They also highlighted Joey's right to be heard as she suggested turning the ovals to vertical instead of horizontal which pushed everyone's thinking about the problem. As they waited, a group member posed a question, "what would be the 2nd best shape to use?" They then then started messing with the stars but noticed rotations weren't eliminating purple space. They also concluded that "teddy bears are the wors" and that "squares would be the best shapes to use."

researcher observational notes—third iteration

The MTE's observational notes from the third iteration exemplify how the PSTs engaged in the notion of group-worthy tasks (e.g., no clear solution nor strategy). The PSTs considered the orientation of the shapes, and which might be another alternative to finding a closer approximation to filling the purple region.

In the second and third iterations of the study (when the PSTs could return to revise their thinking), they returned to the whole group and continued to engage in further self-reflection about their strategy and revision:

Blossom: I guess we just … we went with first one but I'm sure the stars and from seeing the hearts gonna fit into it too. … I knew the bears are not the way to go. So, they just look too big to fit into the purple box so. (Second iteration).

Kacey: And then we did end up kind of looking at other peoples and we decided to see if they would fit on the inside and it kind of solidified what we already had. (Second iteration).

Miller: Because we all got to decide how to solve the problem that made the most sense for us. (Third iteration).

Mallory: And we definitely exercise all of our rights during this but more specifically, the right to be confused a little bit mental game to, you know, the very beginning when what shape are we going to use. How are you going to fill this in? And then the right to make a mistake because initially we thought there would be for ovals in the blank space. But then when we play around with that we discovered, it's more, it's closer to the two ovals as opposed to four. And then also just the right to be heard. Because there towards the end, we were just like playing around with it and kind of getting to, you know, some different conclusions. (Third iteration).

The PSTs considered how they exercised all of their RotL in this group-worthy task when they approached the problem, found a strategy, reconsidered that strategy, and offered a new strategy.

Concluding Thoughts

In this lesson study, MTEs engaged PSTs in the RotL by (1) watching a video clip of a child exercising their RotL while solving a problem, (2) reading an article about the theoretical underpinnings of the RotL, and (3) engaging in a group-worthy problem-solving task of their own. We found through the debriefs that the video clip supported PSTs to consider how the RotL were a fundamental practice that children exercise when they solve mathematics problems and the task built upon the notions of the

RotL. We found that the revisions of the research lesson created more opportunities for PSTs to share more of their thinking by exercising more RotL.

Traditional mathematics classrooms assume that the teacher is the only knowledge holder; the RotL explicitly argues that teachers should honor students' ideas and interpretations as they learn complex ideas. When students' ideas are honored first, students realize the power in their thinking as it naturally arises as opposed to thinking that is filtered through traditional classroom discourses and practices (e.g., repeating algorithms that might not make sense to students at the time). When MTEs create opportunities for PSTs to do more than read about the RotL, but also experience these rights as they are learning mathematics, PSTs have a core experience to rely upon when implementing these RotL in their practice. Through the interactive process of lesson study, the group of MTEs were able to improve the learning opportunity for a deeper and richer experience for the PSTs.

The purposeful questions asked by the MTEs in this study support the existing research about ways to elicit students' mathematical thinking beyond recall knowledge (Eddy & Kuehnert, 2018). By asking "How can you convince me that your answer is the best fit?," PSTs needed to do more than offer an answer. They also had to build an argument as to why their response was the "best" estimate for the area. Since the MTEs were not interested in only a correct answer to the problem, it was important to focus the line of questioning on *how* the PSTs arrived at their answer and *justified* their thinking.

Finally, learning and implementing the RotL is not a process that is quick or straightforward. Rather, teachers and PSTs learning to implement the RotL require self-reflection about how these rights may not have been available to them as students. When PSTs embrace the RotL during math methods, they can find small, yet tangible steps to incorporate them into their emerging practice. By valuing and celebrating students' thinking, more teachers can support children in adopting this thinking for themselves. Without these moments of self-reflection, the RotL risks being another set of static classroom rules.

Implications from our study suggest that lesson study supports PSTs to learn about the RotL in their practice. By revising existing tasks to be group-worthy, PSTs can reflect on the RotL. The RotL are not limited to just mathematics education; science, literacy, social studies, and other teacher educators can also engage in the RotL. The interdisciplinary nature of the RotL is especially important for elementary teachers who already see their work across disciplines. When PSTs are prepared to model for their students the world in which they live, they have more opportunities to use the tools that will support their success and to take on the challenges they might face in the future.

References

Carpenter, T. P., Fennema, E., Franke, M. L., Levi, L., & Empson, S. B. (1999). Children's mathematics. *Cognitively guided instruction*. Heinemann.

Cohen, E. G., & Lotan, R. A. (2014). *Designing groupwork: Strategies for the heterogeneous classroom third edition*. Teachers College Press.

Eddy, C. M., & Kuehnert, E. A. (2018). The advancement of teacher questions in mathematics education. *American Educational History Journal, 45*(1/2), 33–53.

Goffney, I. M., Gutiérrez, R., & Boston, M. (Eds.). (2018). *Rehumanizing mathematics for Black, Indigenous, and Latinx students*. National Council of Teachers of Mathematics

Gupta, S., & Pathak, G. S. (2022). Ethical issues in virtual workplaces: Evidence from an emerging economy. *European Journal of Training and Development* (ahead-of-print). https://doi.org/10.1108/EJTD-03-2022-0023.

Hollins, E. (2011). Teacher preparation for quality teaching. *Journal of Teacher Education, 62*(4), 395–407.

Joseph, N. M., Hailu, M., & Boston, D. (2017). Black Women's and Girls' persistence in the P–20 mathematics pipeline: Two decades of children, youth, and adult education research. *Review of Research in Education, 41*(1), 203–227. https://doi.org/10.3102/0091732x16689045

Kalinec-Craig, C. A. (2014). Examining my window and mirror: A pedagogical reflection from a white mathematics teacher educator about her experiences with immigrant Latina pre-service teachers. *Association of Mexican American Educators Journal*, 8(2), 45–54.

Kalinec-Craig, C., & Robles, R. A. (2020). *When the classroom rules are reimagined as rights of the learner*. Mathematics Teacher: Learning and Teaching Pre-K–12.

Kerckhoff, A. C., & Glennie, E. (1999). The Matthew effect in American education. *Research in Sociology of Education and Socialization, 12*(1), 35–66.

Lewis, C. (2002). *Lesson study: A handbook of teacher-led instructional change*. Research for Better Schools.

Lewis, C., Friedkin, S., Emerson, K., Henn, L., & Goldsmith, L. (2019). How does lesson study work? Toward a theory of lesson study process and impact. In R. Huang, A. Takahashi, J. daPonte (Eds.), *Theory and practice of lesson study in mathematics*. (pp. 13–37). Springer. https://doi.org/10.1007/978-3-030-04031-4_2

National Council of Teachers of Mathematics. (2014). *Principles to actions: Ensuring mathematics success for all*. National Council of Teachers of Mathematics.

Schulte, A. K. (2009). The demographics of teaching and teacher education: The need for transformation. In *Seeking integrity in teacher education* (pp. 23–31). Springer.

Smith, M. & Stein, M. K. (2018). *5 Practices for orchestrating productive mathematics discussion*. National Council of Teachers of Mathematics.

Torres, O. (2020). *Equity in Education Webinar Series: Rehumanizing Schools - Rights of the Learner*. Retrieved September 9, 2020, from https://www.youtube.com/watch?v=_UndpNUCAqw&t=3211s

Additional Examples of Lesson Study in Preservice Teacher Education

12

USING LESSON STUDY TO BUILD INTERDISCIPLINARY STEM COLLABORATIONS AT AN URBAN COMMUTER UNIVERSITY

Janelle M. Johnson, Nursen Konuk, and Mark Koester

Preservice teachers (PSTs) of secondary mathematics and science are typically prepared to teach their disciplines separately (Brown & Bogiages, 2019). However, with the growing popularity of problem-based learning in STEM-based schools and the broad recognition of the importance of data literacy, teachers need to think outside disciplinary silos. While we can see examples of interdisciplinary teaching and learning under the Career and Technical Education (CTE) umbrella, we rarely observe interdisciplinary learning opportunities in the so-called "college prep" track. We therefore see an urgent need for greater linkages between mathematics and science if we hope to close opportunity gaps in STEM and beyond, and support the development of problem-solvers who can address highly complex challenges (Miller, 2017). Our approach, based on our years of combined teaching practice at K-12 and university levels, is interconnected with theories in the research literature. However, in our state and institution, there are longstanding structural barriers to interdisciplinary collaboration:

- The siloed nature of mathematics and science instruction at both secondary and tertiary levels (Lesseig et al., 2016; Neilson & Campbell, 2018);
- The nature of secondary mathematics and science teacher licensure at the state level; and
- The lack of space and time for co-teaching at all levels of our learning ecosystem.

Furthermore, while we highly value the benefits of arranging PSTs' clinical experience in diverse high-needs schools, we have had challenges with school partnerships due to teacher and administrator turnover, lagging test scores that often contribute to reluctance to host PSTs, and overstretched personnel in both the K-12 schools and the university. In this chapter, we reflect on how lesson study

DOI: 10.4324/9781003326434-17

with PSTs has offered a promising starting point for interdisciplinary teaching and learning despite these challenges. Creating structures for real life and relevant problem-solving in mathematics and science-based educational spaces offers much transformative potential for closing opportunity gaps in STEM fields. In this chapter, we describe the context of our study and offer some background on the experiences with lesson study at our university. We then outline our research design, followed by our preliminary findings and some concluding discussion.

Study Context

At our university, secondary PSTs major in their content area and take a sequence of lecture and clinical field courses in the School of Education to fulfill their concentration for licensure. They are encouraged to see advisors in both their major and within the education department, but that occurs to varying degrees since some institutional discourse and expectations need to be more transparent to students. Each department that houses the major has autonomy in PSTs' course of study, but students sometimes have difficulty accessing accurate information about their required coursework, depending on the department. Some departments have built extensive course offerings to meet the needs of PSTs in their content area. The mathematics education program at our institution is one such department and has had an established lesson study component for about 15 years. When secondary PSTs are in their mathematics methods course, they participate in two lesson study cycles through a mathematics content course for elementary PSTs that they co-teach.

In contrast, prior to this study, our science teacher preparation program had no history of lesson study, at least partly due to the absence of a science education department. Unlike all other areas of teacher licensure, students take only one science methods class before student teaching, and the course content tends to feel quite compressed and urgent to both students and the instructor. As Marble (2007) writes, "There simply is not enough time in one preservice course to transmit much more than the simplest, most mechanistic vision of practice" (p. 936). Another difference between the two departments is that the math department has more autonomy over PSTs' clinical placements, while the science PSTs are placed by a central office.

Secondary science teacher licensure in our state, as implemented in our accredited institution, requires students to major in Environmental Science, Biology, Chemistry, or Physics and take coursework and labs in the three other areas. (Postbaccalaureate students similarly name a content focus and complete coursework parallel to the undergraduate program.) Each major department historically had little to no communication with the three other science departments in determining the course of study for licensure students. As reported by our colleagues, the requirements for teacher licensure were generally not considered when designing the majors. Though our university has made some progress in this area, the number of credit hours that this breadth of coursework requires means that science PSTs often butt up against credit hour caps. They would also have little to no room for additional math-based coursework or experiences to potentially improve their data literacy. Students frequently reported how siloed each science course was across the sciences, leaving them somewhat unsure how to reconcile those conflicts, especially in their roles as future teachers. While the integrated type of STEM we see in some PK-12 schools and districts is promising for interdisciplinary teaching and learning, we, unfortunately, do not yet offer that pathway to PSTs. These factors were all motivators for the study we describe here. We hope that the collection of research-based evidence will help bolster our case for making some much needed systemic shifts.

Background on Lesson Study with PSTs at Our Institution

I (Janelle) am our university's sole secondary science methods instructor. In Spring 2019, I experimented with informally engaging the three science methods students in my class in lesson study through collaboration with a mathematics methods instructor. The three science PSTs adapted and conducted one cycle of lesson study (i.e., study/plan/teach/reflect) in the elementary mathematics course with a tree farm problem I selected from a Common Core State Standards (CCSS) math assessment resource; I felt the science PSTs would readily see the relevance of a tree farm problem that they could potentially include in a biology or environmental science course, for example. The problem included a diagram of a large tree farm with symbols representing different ages of trees, and blank spaces where no tree is present. The diagram contains too many symbols for teacher candidates to realistically be able to count them, so they have to practice estimation strategies, first individually and then by coming to consensus with other group members. The problem asks learners to articulate their methods to the rest of the class, then reflect on their work and what they learned during the process. PSTs planned the lesson and then divided into pairs to co-teach in a section of the elementary math methods class. After the teaching, during the next methods class, the group debriefed with reflections from the co-teachers and evidence collected by four to five observers who were fellow students in the math-science methods group. The science PSTs were excited and pleased with their lesson study experience and in their summative course assessments recommended that more lesson study experiences be offered to science PSTs.

In the next semester, Janelle and Nursen, along with another colleague, decided to teach a combined science and math methods class. I (Janelle) was able to get the science methods course's schedule aligned with the mathematics methods course timing, but limited classroom space was another challenge. A plausible solution arose when we planned to combine the mathematics and science methods courses into one, and everyone could pack into the same classroom. All secondary PSTs had previously taken a general lesson planning, assessment, and management course before the combined math-science methods course, and most were also taking a disciplinary literacy course that same semester. Our ability to do adequate co-planning was limited due to the ad-hoc nature of the collaboration and the short timeframe. Since our main goal as instructors was to engage the PSTs in lesson study, we utilized the existing well-established math methods class syllabus centered on lesson study as a basis for the combined math science course.

Unfortunately, most lessons historically used as a basis for the lesson study in the math methods course were not interdisciplinary and offered seemingly little relevance to the science PSTs. The first lesson study task was a card probability problem. There were substantial differences in mathematics understanding and mathematical problem-solving ability between the science and mathematics PSTs. Several of the science PSTs expressed a notable need for more confidence in their mathematics ability, let alone to teach it comfortably. We will share some learnings about how to develop a class culture that values both mathematical and scientific knowledge later in the chapter. The class ratio was another factor in the group's dynamics since the math students heavily outnumbered the science students. Another difference between the two groups was that the math PSTs had taken a math-specific Curriculum Standards Course that does not exist for the science PSTs.

A couple of the science PSTs exhibited anxiety as they struggled with the mathematics in the lesson study steps. Some science students began to vocalize that they "weren't math people" and began to disengage from the course overall. (It is important to note that the science students were also provided opportunities to do lesson study in some university science courses, and the same few students demonstrated notable low self-efficacy in teaching sciences outside their major.) As the science methods instructor, Janelle tried to inject more interdisciplinary content into the lessons with the hope that both groups would see the relevance of the lesson topic, so we decided to replace one of the more traditional math-based lessons with a fish sampling problem. During the study step, we provided all the PSTs with local current events articles about fish sampling to provide more context and application for the problem. Though a routine practice for the science PSTs, the mathematics PSTs seemed to struggle with the use of current events in their mathematics instruction. The lesson study groups that were all numerically dominated by math PSTs dropped the current events article from the lesson. This was evidence that we as teacher educators had not been explicit enough with all the PSTs about the importance of building lessons relevance to learners through such components as current events articles (Lambert et al., 2018).

Research Design

The study described here explores how a pair of novice mathematics and science teachers co-planned lessons that integrated mathematics and science concepts for an interdisciplinary mathematics and science class in a middle school. We reflect on how teachers who had been PSTs in our teacher preparation program drew on their experiences and dispositions developed during lesson study to implement interdisciplinary teaching and learning. To search for interdisciplinary connections, we utilized qualitative analysis of artifacts, including drafts and revisions of the co-created lesson plans, photo documentation of interdisciplinary curricular planning sessions, transcriptions of audiotapes of planning sessions, and teacher interviews.

During the following semester in Spring 2020, one of the math PSTs[1] who had been in the combined methods class was a student teaching at a local, small early college serving a highly diverse population. A staffing challenge at the school as COVID-19 hit meant that she and another math PST were co-teaching high school mathematics, an experience they enjoyed and described benefitting from. These two teachers had participated in a lesson study cycle during their student teaching through the math department, and they both conveyed that it had been very successful and helped them feel comfortable receiving critical feedback; their professionalism and flexibility contributed to the degree to which the school principal positively viewed the teaching abilities of PSTs from our university and actively sought to hire students from our program.

As the new school year began in the fall of 2020, COVID-19 restrictions meant that secondary students at the early college who had four core classes could only see two teachers in a day. The school's solution was to pair middle school teachers to combine in-person and remote instruction. The former math PST mentioned above, who is now a teacher there, partnered with a science teacher, who was another graduate of our program. They began to innovate how to combine their two classes more effectively and address some challenges around student self-efficacy they were observing. They mapped out each grade level concept from the standards and began to intentionally align their instruction. The two teachers reached out to the study authors with great excitement about what they were observing. They saw substantial improvements in student engagement and performance and increased self-efficacy among previously struggling students. They described explicit connections between related ideas in mathematics and science and explained that lesson planning between the two of them was easier because of the co-planning during the teacher preparation program. We were excited by the unique possibilities this situation offered and decided to construct a small research study to examine it.

Preliminary Findings

Their undergraduate lesson study experience helped the teachers discover that specific study topics lent themselves better to interdisciplinary teaching. These early career teachers' research on key documents during the *study* phase helped to design lessons that authentically used mathematics and science. In their 8th-grade co-teaching unit, for example, weather and space were the topics most fruitful for lesson study (Figure 12.1). Students maintained weather journals and created linear and nonlinear graphs with their collected data. The teachers described how the nature of the unit facilitated student understanding of math concepts and terminology. The teachers' lesson study connected the science standards about space and planets with the math standards of area and circumference of circles and surface area and volume of spheres. The math teacher described how lessons based on space helped students "discover" pi with an engaging lesson and lay a better foundation for their understanding of volume.

I started teaching circles and spheres. He has taken it to planets, and they are doing scientific notation, substitution into formulas, that maybe they haven't

Science helps Math

FIGURE 12.1   Student's "Bad Drawing" Showing Connections between Science and Math.
(*Continued*)

A Symbiotic Relationship
Math helps Science

FIGURE 12.1   (Continued)

experienced everything with yet. They have a basic idea of what scientific notation does, but they know at least how to plug and check some data. They know
what the sphere is doing, and they recognize patterns, because we discovered pi.
We discovered how area works by cutting up a circle. We went through all those
discovery pieces, and they recognize even in volume they haven't directly studied in math yet … they're headed there. (The science teacher) actually ended up
going fast enough, he's actually going to foreshadow my math. And so we end
up side-by-side teaching content this semester with much more purpose.

The teachers were thrilled with the degree of student learning they observed due to the interdisciplinary lessons. They were especially pleased with the shift they observed in a student they had "struggled to reach for a year and a half," describing the student's excitement as he shouted, "I've got it!" and completed work that before he would not have even started. The math teacher described how a colleague who teaches in the high school, "says her kids kick and scream, 'Why am I having to do math in the middle of science?' Our kids go, 'Of course it ties. You need math for science,' and it's a different mindset." Students do not automatically see the connection between math and science because they are taught separately. The math teacher continues:

> This approach builds cross-curricular bridges that encourage the content to be more accessible to students. We're creating great connections, and the kids are starting to use math and science vernacular naturally.

The teachers recognized that their own experiences of learning math and science in silos during most of their teacher preparation coursework were not the only limitation to their interdisciplinary efforts. The students themselves had only thus far experienced math and science learning separately, and as the teachers reflected on the students' feedback, realized that they needed to make the interdisciplinary connections explicit. During the co-taught class, the teachers built a lesson around Chapter 4 of Orlin's (2018) book, Math with Bad Drawings. Chapter 4 is called "How Science and Math see Each Other." Students were asked to read the chapter and create their own "bad drawings" to illustrate the friendship between science and math (Figure 12.1). The need to make such explicit connections is an important consideration for teacher educators and instructional coaches.

In addition to the improved and applied student learning the teachers observed, they described how the interdisciplinary lesson study helped them address gaps in their own knowledge base and teaching experience, echoing the findings of other research about the benefits of integrated STEM learning (Brown & Bogiages, 2019; Lambert et al., 2018). Their co-teaching helped them pinpoint what tools and graphic organizers would best help them meet their student learning objectives. They describe their professional growth that resulted from the lesson study cycle to situate their content area in connection with other content areas. The collaboration motivated them to seek out interdisciplinary curricular materials, but they described their frustration and disappointment when they could not locate any such resources.

> What I am getting is from working side by side on the lesson study, and so if we can build purposefully, to look at how we can make more organic experiences, where I talk about math *and* science. We could backward-plan a lesson set and talk about, "Okay, here's your two topics you're going to teach. How do you make them touch?" Thus, even if it is simply significant figures and place value,

knowing that math works to tie it to that in elementary math. We cannot assume they understand that. That is where the problems start, many times in math, from the elementary years when they did not comprehend it, and then it comes up here, and you are trying to put out that fire.

The co-teachers' interdisciplinary lesson helped elicit reflections about their teacher preparation program and its limitations. They realized that the math PSTs could benefit from tapping into science topics as a basis for math learning and that many of the science PSTs would benefit from a better understanding of math concepts. These teachers' interdisciplinary lesson study helped them realize the benefits of co-teaching, co-planning, and increasing communication with colleagues in different departments.

Concluding Thoughts

The middle school teachers at this school unknowingly laid a foundation for the lesson study collaboration when their offices were combined, and they got a glimpse at the standards the other teachers were planning lessons around. This clarity generated conversations around possible connections across the content areas. Schools that want to offer interdisciplinary learning opportunities for students need teachers who have had opportunities to learn how to connect curriculum between and across traditionally siloed content (Lambert et al., 2018). Lesseig et al. (2016) write that "negotiating the [interdisciplinary] implementation within the constraints of traditional school settings was daunting, particularly when these kinds of projects were new to both teachers and students" (p. 182); their analysis revealed pedagogical, curricular, and structural challenges. Our preliminary analysis suggests that extensive revision is needed in our coursework and field experiences if we hope to facilitate teachers' capacities for connecting mathematics and science learning. Typically, the *study* step was not a prominent part of our lesson study, but we see that we need to have a more robust *study* step for PSTs to be able to make connections between disciplines. How can we better blend the needs of math and science PSTs in a combined methods course that includes lesson study? We now know that more work is needed to create a sense of belonging in the combined methods class, including valuing the science PSTs' content knowledge. What elements of the lesson study cycle are most beneficial to the teachers in the current ecosystem and constitute effective professional development (Rogers et al., 2007? The *study* and *plan* steps need to be expanded in the combined methods class.

While the interdisciplinary lesson study we attempted during the PSTs' combined methods class seemed largely unsuccessful at the time, that experience planted seeds in the PSTs' minds about interdisciplinary co-teaching and learning. For these two teachers, at least, the outcomes of the experience were positive both for their professional growth and for their students' learning and mindsets. Our findings are consistent with the research about integrated STEM learning and the preservice lesson study we reviewed. This work helps us understand how to construct better lessons for lesson study that integrate both math and science ideas so that all PSTs see more application and relevance to their teaching (Beauford, 2009; McHugh et al., 2017). They will allow us to backward design our coursework and clinical field experiences. We also hope these findings can build toward contributing to state-wide discussions of teacher licensure and help us revisit our university structures for secondary teacher preparation. Ideally, this is the beginning of much longer-term work that can help us better support PSTs and novice teachers in feeling comfortable with interdisciplinary instruction that closes opportunity gaps for their students and contributes to the larger conversation on lesson study as pedagogy. It offers questions about models of teacher preparation and licensure, the nature of standards and course offerings, planning time and resources for teachers, how faculty and teacher workspaces are structured, and the benefits of co-teaching and connecting with a fuller range of students.

Acknowledgment

We gratefully acknowledge the support of National Science Foundation Robert Noyce STEM Teacher Scholarship funding under Grant #s 1540805, 1660506, & 2151027.

Note

1   The math PST was also a Noyce Scholar. All three of the authors were members of the Noyce leadership team at the time this chapter was written. Because of this relationship, the researchers had more regular communication with the PSTs as they transitioned into teaching.

References

Beauford, J. E. (2009). The great divide: How mathematics is perceived by students in math and science classrooms. *Science Scope, 33*(3), 44–48.

Brown, R. E., & Bogiages, C. A. (2019). Professional development through STEM integration: How early career math and science teachers respond to experiencing integrated STEM tasks. *International Journal of Science and Mathematics Education, 17*(1), 111–128.

Lambert, J., Cioc, C., Cioc, S., & Sandt, D. (2018). Making connections: Evaluation of a professional development program for teachers focused on STEM integration. *Journal of STEM Teacher Education, 53*(1), 2.

Lesseig, K., Nelson, T. H., Slavit, D., & Seidel, R. A. (2016). Supporting middle school teachers' implementation of STEM design challenges. *School Science and Mathematics, 116*(4), 177–188.

Marble, S. (2007). Inquiring into teaching: Lesson study in elementary science methods. *Journal of Science Teacher Education, 18*(6), 935–953.

McHugh, L., Kelly, A. M., & Burghardt, M. D. (2017). Teaching thermal energy concepts in a middle school mathematics-infused science curriculum. *Science Scope, 41*(1), 43.

Miller, R. K. (2017). Building on math and science: The new essential skills for the 21st-century engineer. *Research-Technology Management*, *60*(1), 53–56. 10.1080/08956308. 2017.1255058.

Neilson, D., & Campbell, T. (2018). Adding math to science: Mathematical and computational thinking help science students make sense of real-world phenomena. *Science Teacher (Normal, Ill.)*, *86*(3), 26–32.

Orlin, B. (2018). *Math with bad drawings: Illuminating the ideas that shape our reality* (pp. 30–37). Hachette UK.

Rogers, M. P., Abell, S., Lannin, J., Wang, C. Y., Musikul, K., Barker, D., & Dingman, S. (2007). Effective professional development in science and mathematics education: Teachers' and facilitators' views. *International Journal of Science and Mathematics Education*, *5*(3), 507–532.

13

THE USE OF LESSON STUDY IN THE ONLINE TEACHING ENVIRONMENT TO SUPPORT PRESERVICE SCIENCE TEACHER LEARNING

Rita Hagevik and Irina Falls

Historically, busy schedules and logistical considerations have made it difficult for teachers, especially in the US, to coordinate and collaborate on a regular basis as required by lesson study practice. New collaborative modalities, over the last few decades, have made communication and collaboration among teachers easier, especially through the rapid evolution of digital technologies. Similarly, teacher preparation programs are constantly searching for new, innovative, time-saving modalities to prepare licensed teachers, given the shortage of teachers across the US. Thirty-two states have alternative licensure programs with a 60% increase in enrollment over the last 10 years (25% of all programs), and the greatest shortages of teachers are in secondary mathematics and science (Yin & Partelow, 2020). At the same time, digital strategies and tools are becoming necessary not only for online learning but also for becoming a successful teacher. Technology-based strategies provide new and promising opportunities to incorporate lesson study into online synchronous and asynchronous learning, both during teachers' preservice preparation and during in-service professional development. We used lesson study in an online instructional format for science education teacher candidates. Lesson study was an added opportunity for students to systematically examine pedagogical practice to become more effective teachers and attain the required digital competency standards for teachers. This chapter explains the use of technology tools to support collaboration and deepen reflection in an alternative secondary science teacher initial licensure program called a Master of Arts in Science Teaching (MAT).

Lesson Study and Technology Tools

Remote learning formats increase collaboration and communication among others and across locations (Weaver et al., 2021). In online lesson study, identifying practices that are purposeful and intentional is paramount. In transferring to an online

DOI: 10.4324/9781003326434-18

lesson study, we maintained the same features and steps as the face-to-face lesson study: *prepare*, *study*, *plan*, *teach*, and *reflect*. We illustrate ways to use digital tools to implement lesson study.

Methodology

We used a case study research approach (Creswell, 1998) to explore how four preservice teachers (PSTs) used the five steps (*prepare*, *study*, *plan*, *teach*, and *reflect*) of the lesson study approach in a fully remote environment. The first author was the instructor of the course and the second author served as an outside observer. Both authors taught online for over 10 years, completed three different online certifications, and taught in fully online programs. The research questions were (1) what are the characteristics of online lesson study in PST education?, (2) what are PSTs' perceptions of knowledge and self-efficacy about science teaching after online lesson study?, and (3) to what extent were PSTs able to reflect upon changes in their teaching practices because of the online lesson study experience?

Participants

The four PSTs called Tina, Jackie, Kay, and Ralph (pseudonyms) were enrolled in a pre-internship course as a part of a year-long internship experience. All four PSTs had undergraduate science degrees. Even though each student taught in a different school system spread throughout the state, they were paired and worked together as two lesson study groups. Group 1 was the middle school group of Tina and Jackie and Group 2 was the high school group of Kay and Ralph, each designing one mini unit of 5 days. Each PST had a school-based mentor teacher for lesson study support. The groups worked parallel to each other during the 3-hour synchronous online weekly sessions for the 15-week course. The PSTs, ages 25–45, were the teachers of record and had 1 and a half to 2 years of teaching experience in high-needs schools, 100% of the students received free and reduced lunch.

Lesson study in the Online Context

The PSTs composed lessons using the Next Generation Science Standards (NGSS) around science phenomena (National Research Council, 2013). They worked with the course instructor and school-based mentor teachers to design the lessons using NGSS-aligned curricula. Middle school Group 1 of Tina and Jackie designed a mini unit called "take a deep breath" that covered general functions of the major systems of the human body (NGSS-MSLS1-2 and LS1-3), the science and engineering practice of developing and using models, and the cross-cutting concept of structure and function. The mini unit focused on how oxygen is circulated throughout the body. The teaching goal of this unit was to increase independent thinking around scientific phenomena. Tina taught the 5-day unit with support from Jackie.

High school Group 2 of Kay and Ralph designed a mini unit called "energy transfer in chemical reactions" that covered the release or absorption of energy in chemical processes (NGSS-HSPS-1.A and B), the science and engineering practice of developing and using models, and the cross-cutting concept of energy and matter. The 5-day mini unit focused on making models and testing chemical processes to explain energy transformations. Ralph taught the lessons and Kay supported him. All lessons were videotaped. The teaching goal of this unit was to build confidence in students and for them to take responsibility for their own learning. The PSTs employed a variety of teaching strategies in the units by teaching the three dimensions of NGSS: Practices, cross-cutting concepts, and core ideas. They used scientific argumentation, graphic organizers (e.g., KLEWS charts), modeling, videos, case studies, labs, lab reports, and data collection and analysis. Through the prepare, study, and plan phases, the PSTs located and modified for their own purposes the lessons and teaching strategies from existing curricula such as NextGen Navigator by NSTA, Ambitious Science Teaching (Windschitl et al., 2020, Argument-Driven Inquiry, and the Argumentation Toolbox.

Procedure

The steps of the online lesson study included:

Prepare (Early August)

1 Learn about the lesson study process and determine group norms and teaching goals using learning management system (LMS) assignments, ZOOM, and breakout rooms.
2 Individual reflection on group norms and collaboration using the LMS discussion board. Adjust group norms as needed during the ZOOM meeting discussions.

Study (Mid-August and Early September)

3 Brainstorm research-based lesson ideas using Jamboard during ZOOM class meetings. Breakout rooms to discuss. Use the chat function of ZOOM for comments. PST discussed ideas with school-based mentors outside of class time. LMS as a resource for lesson planning.
4 Use of ATLAS videos of best practices in teaching (atlas.nbpts.org) to demonstrate research-based teaching strategies through discussions.

Plan (Mid and Late September)

5 Develop research-based lessons and supplemental materials collaboratively using Google Drive and Google docs.
6 Peer review lessons through comments collaboratively in Google Drive. School-based mentor teachers comment.

7 Reflection and revision of lessons and instructional materials using Google Drive. Presentation of the research-based lessons and discussions in ZOOM.

Teach (October)

8 PSTs choose one person to teach the five lessons and to record each lesson and the teaching videos are uploaded to Google Drive. PSTs use iPads and other devices to ensure sound quality. GoReact offers PSTs support for creating high-quality teaching videos.
9 PSTs, the instructor, and mentor teachers watch the teaching videos and discuss them with PSTs. During class via Zoom using the whole class and breakout rooms, the class discusses and reflects on the evidence during the debriefing session(s).

Reflect (November)

10 Watch party in which each person selects their favorite teaching lesson and uses GoReact to create lesson tags/markers. Zoom discussion on the high-impact to lower-impact student engagement strategies.
11 Reflect using the discussion board on the LMS system on the teaching debriefing and individually on the process of lesson study.
12 Discuss ways the lesson could be changed when taught the next time and reasons why explored. Final reflection on the lesson study process and complete the lesson study survey for instructor feedback in Qualtrics.

The teachers of the lessons obtained videotaping permission letters that were signed by the students and the parents. Artifacts such as student work, grading rubrics, videos, and lesson plans were shared on the Google Drive established for the course. Student work was given pseudonyms (numbers for example) to maintain anonymity.

Data Collection and Analysis

We used multiple data sources which included: (a) Written mini unit and supplementary materials; (b) teaching videos and notes; (c) debriefing sessions; (d) discussion board reflections; (e) watch party and debriefing; and (e) individual PSTs final reflection. We used a rubric by Ward and McCotter (2004) to analyze the depth to which the PSTs reflected upon changes in teaching practice.

The data sources were read and discussed by the first and second authors to develop the themes to answer the research questions. The debriefing sessions, discussion boards, and reflections were analyzed using the Ward and McCotter (2004) rubric to determine how the PSTs' views of teaching related to student learning changed over the semester. We examined the ability of the PSTs to use reflection on their own practice to improve through a course of action or future responses. Ward and McCotter's (2004) rubric identified three areas of reflection emphasis:

Focus, the process of inquiry, and change in practice and perspective at four levels: Routine, technical, dialogic, and transformative. Beginning teachers are not expected to reach the transformative level because of inexperience. For PSTs, the goal is to operate at the dialogic level where they situate questions and actions by considering other perspectives and gaining new insights as a result. The instructor of the online lesson study used this reflection rubric throughout the five stages of the lesson study to gauge the PSTs' growth through their reflections throughout the 15-week course. Doing this facilitated individual support for each PST as they progressed through the lesson study steps in the online environment.

Findings

RQ1: What are the characteristics of online lesson study in PST education?

Prepare

We used online resources on the process of lesson study through the CANVAS LMS to teach the PSTs the process of lesson study. This was followed by online discussions using Zoom during class meetings. Choosing groups and establishing norms were accomplished through discussion board assignments once consensus was achieved. Group breakout rooms during the class time were used to establish PSTs' meetings outside of class in the online environment through Zoom. Online lesson study facilitates this process by establishing an electronic flow of information and communication between the PSTs and the on-site mentor teachers using a shared Google Drive. PSTs and their mentors established common rules on how to access and use the online resources included in the CANVAS course and on Google Drive. Google Drive was used as a repository of research documents and as a platform for collaboration on the same document when needed.

Study

The PSTs used high-quality online resources to organize and build PSTs' knowledge of the topics. Videos about good teaching (ATLAS–National Board of Professional Teaching Standards as well as other online collections) were used together with videos of the process of lesson study to demonstrate instructional strategies and to access the NGSS-aligned curriculum. All these videos were discussed in virtual meetings facilitated by the two instructors (university mentors), while specific instructional strategies were discussed and demonstrated during virtual class meetings. PSTs discussed their choice of curricular topics and ideas with each other and the school mentors, after which they were presented during our class meeting.

Plan

Digital learning tools focused the PSTs on planning. The students discussed their lesson plans during virtual meetings facilitated by at least one of the authors. They developed their plan using Google Drive tools for co-writing and collaboration.

Once created, the faculty mentors used online editing tools as well as peer review and comment to offer suggestions and revisions. Online editing tools offer an opportunity for asynchronous continuous improvement and discussion. In addition, discussion boards through CANVAS were used to reflect upon and offer ideas during the planning stage. The PSTs discussed first as a group and made planning decisions, after which a session with the instructor was held. Reflective questions on the lesson plans were offered by the instructor of the course using the lesson study format through the discussion board. The focus of lesson study is on student learning, so PSTs were challenged to connect the instructional strategies to student outcomes and to their identified teaching goal. For example, Tina taught with Jackie's support since Tina taught 7th grade in a rural setting (K-8 school) and Jackie taught 6th grade in an inner-city middle school (grades 6–8), so the standards were different. They met via Zoom on evenings and weekends and communicated via email in between. They used a Google calendar with reminders and the course Google Drive to store and share documents. For each lesson plan, they identified student behaviors to be identified during the viewing of the teaching videos. However, their overarching teaching goal was for students to use independent thinking to solve problems. They designed a unit on the respiratory and circulatory systems in which the students would figure out how oxygen circulated through the body. Tina noted during an online class that she had several students with limited English proficiency, so the use of pictures and models was important to support science vocabulary acquisition. Tina stated, "collaborating and planning online showed [her] how to work collaboratively with [her] own students."

Teach

Digital learning tools promoted discussion. PSTs videorecorded the five lessons in each of the mini units. The teaching recordings were uploaded to Google Drive. Each PST group watched the teaching videos and used GoReact to create the tags asynchronously. The instructor taught the PTSs how to create markers in GoReact to focus their attention on student behaviors/outcomes. Each PST group selected one teaching video for the watch party completed during class. Each group selected the video that represented the "best" teaching in their view in relation to NGSS and student learning. During the watch party, each group presented a summary of what they learned and introduced the teaching video selected and why they selected it. The PSTs were placed in breakout rooms and watched the other group's teaching video, using GoReact to create the tags. Then each group shared with each other what they learned. A discussion board followed as an assignment in CANVAS. Teaching videos were discussed within groups and among groups. An advantage of using GoReact to analyze teaching videos is that you can comment at a certain time stamp of the video and respond to other people's comments. Due to the many features of GoReact, PSTs offered each other substantive and precise feedback regarding the teaching videos. This strategy was modeled for students using the ATLAS videos during the plan stage. Tina and Jackie reported during the online debriefing session using GoReact and tags that they noticed that there were times when "the students did not have anything to do, and I really have noticed some techniques I need to work on after watching myself teach." Ralph showed support during the online discussion by remarking, "pacing is hard." Tina said,

> Transitions are something I need to work on. The students did well with the lungs, but the heart was harder for them to understand. It would have been better for the students to create a model of the heart before discussing and teaching the structure and functions.

The digital tools supported PSTs' discussions through observations of their students' talk. The debriefing and close analysis allowed for a deeper understanding for all the PSTs.

Digital learning tools promoted debriefing on teaching and focused PSTs on students' learning. Ralph and Kay, the high school group, noted that after the experiment, "Students were able to organize their arguments and use evidence to connect their claims. Students need to practice their reasoning skills more by talking through their thoughts out loud in class. Students had trouble connecting claims to evidence properly." Through online discussions and using GoReact to collect evidence, critical reflections occurred such as when Kay observed, "Students had difficulty critiquing each other's posters. They did not know what type of feedback to give and how to have the eye required to pick out things in each other's writings and posters." Ronnie then offered a solution, "Students need more instruction on what is necessary for these assignments. Giving feedback requires a strong understanding of what it takes to do well on the assignment." The PSTs found success in using a variety of resources such as school-based mentors as others, online discussions, record keeping in Google Drive, digital recording of lessons, reflection through discussion boards, using GoReact to observe and collect data, and collaboration through breakout rooms and chat to be critical and supportive of teaching and student learning.

RQ2 and RQ3: How does online lesson study support PSTs' perceptions of knowledge and self-efficacy, and to what extent were PSTs able to reflect upon changes to practice?

Teach

The PSTs agreed that they learned new instructional approaches and tools for teaching science using online lesson study. Modeling of instructional approaches occurred in multiple formats facilitated in the online lesson study environment through video critiquing, high-quality resources organized in CANVAS, use of video analysis, examples in Google Drive, and communications through discussion boards or synchronously in breakout rooms, for example. GoReact focused PSTs on student behavior that demonstrated learning based on predefined research teaching goals. Multiple observers and using the video segments as evidence to explain targeted learning behaviors generated rich and productive discussions. Jackie said, "Adapting a student-centered approach allows students to become more independent in their thinking. Scaffolding is important in assisting students with discovering their answers." Kay realized, "I need to give my students more time to practice what they are learning." Jackie agreed when she said,

Some of the students needed additional time for each lesson, more in-depth instruction to get a better concept of the material. A mixture of different learning styles in a classroom is a huge challenge, because of how to help a student that needs additional instruction without losing the student that can understand concepts quickly. I attempted to reduce this problem with a mixture of groups.

Ralph remarked,

the students needed more practice, especially in writing scientific arguments. This could be done separately from the assignment as an activity in the classroom. This way students would have a better idea of how to identify what claims should be supported by what type of evidence.

Kay said, "I agree with Ralph in that my students needed more language support during the lessons." The PSTs agreed that their students needed "more one-on-one attention and individualized help." Tina offered a suggestion, "I have learned new strategies that I can use for my struggling students to better understand the concept such as continuous assessment rather than just testing." Ralph summarized by saying, "We are beginning teachers and we now have a better understanding of how we can actually conceptualize how student learning will happen." Effective use of teaching videos promoted PSTs' critical discussions and analysis leading to deeper understanding and learning.

Reflect
The PSTs gained confidence through online lesson study by identifying areas of improvement. Kay said and Jackie agreed that "while watching the videos I saw things that my students could benefit from if I would add more explanation to assignments and ask more probing questions when students present ideas or projects." Ralph said,

I need to ask more probing questions when students present ideas or projects. Asking students probing questions will help them think on a deeper level and bring their ideas to life. By adding more explanations for assignments and activities students will be overall more confident with the task.

Tina further explained, "I want to use more real-life examples. Creating more opportunities for students to connect their cultural experiences to the content learned in class will enhance student learning." Ralph said, "I am learning how to use something as complex as scientific argumentation […] with my students." Jackie said, "I am less likely to give students worksheets." Tina said, "I am more likely to collaborate with my co-workers on learning strategies." Kay said, "Lesson study showed me the importance of reviewing my students' work and progress. So that I can help them better to understand certain concepts in the future." Ralph said, "I also understand that students can do an assessment to demonstrate understanding, but that there is also value in asking students for feedback in general, especially on their own feedback (peer review process)." The PSTs gained a new understanding of planning as Ralph continued, "I am learning that there is much more to planning than just having a list of activities to do. It is better to over-plan than to under-plan. I should have a more holistic understanding of the lesson that I am teaching and not take the process one step at a time to plan better, I should always be looking ahead." Tina said, "I learned that I need to spend more time on lesson planning. I need to think through my lesson plans more thoroughly." Jackie remarked, "I learned that small changes in my lessons can equal big results for my students. I want to experiment with other teaching tools to improve my teaching." The PSTs described the process of online lesson study as "exhausting but educational, long and tiring, and complex." Ralph clarified by saying, "Lesson study is a great way to break down the planning, teaching, and assessing process by having us carefully do each and reflect upon them as we do them." Through the online lesson study process, PSTs saw, learned, practiced, and observed the importance of planning and the connection to students learning.

PSTs agreed that besides collaboration, the most important part of the online lesson study *experience was reflection.* The high school group of Kay and Ralph began at a *routine* level of focus, inquiry, and change according to the reflection rubric by Ward and McCotter (2004). The middle school group of Jackie and Tina expressed a more open approach to the specific teaching tasks and began at a *technical* level of focus, inquiry, and change. By the end of the online lesson study, the PSTs moved toward a more sophisticated *dialogic* view of teaching based on their focus, inquiry, and change and listened to the perspectives of their students, peers, and others. They identified changes to their teaching practices, for example, asking probing questions, providing better instructions, and more individualized feedback to students. The PSTs found value in collaboration as evidenced by these selected quotes from all four PSTs,

> I like the way the lesson study group planned the lesson as a whole. This allowed each member to introduce methods and strategies to enhance student learning. We worked together for continuous improvement. It was helpful to have someone to bounce ideas off.

The online lesson study environment fosters a continuous and dynamic exchange of ideas between multiple groups over an extended period, not restricted to a meeting time, and even when these individuals are in different places. Overall, the PSTs moved toward a more critical form of reflection by developing new insights about their own teaching as well as about student learning.

Implications

These results point to several opportunities and challenges when using online lesson study. These results showed that these PSTs who went through only one cycle of online lesson study, without prior knowledge of lesson study, were able to perceive the changes that needed to happen in their teaching through analysis and reflection. This was facilitated by having all documents and recordings digitally organized and shared in one place for all group participants to access at any time. The PSTs, with guidance, became their own observers and analyzed their own teaching and as a result, came to new meanings through the recognition of the true challenges facing their students (Larssen et al., 2018). The PSTs learned in an online, safe, well-structured, and supported environment. They learned to shift the focus from instructional activities and materials (technical) to the effects these have on student learning (dialogic). The online environment facilitated a collaborative public forum that led to greater opportunities for questioning the personal beliefs and practices of the PSTs as they taught (Weaver et al., 2021).

Concluding Thoughts

The use of online lesson study offered a unique opportunity to overcome barriers PSTs encounter when entering the teaching profession. The online environment afforded the opportunity to include others such as school-based mentors in the sharing of documents, videos, and student artifacts during the process. Collaboration and collective discourse were encouraged not only within the lesson study but between the groups. The online environment encouraged collective discourse on content, pedagogy, teaching practices, and student learning. The fact that everyone could contribute synchronously and asynchronously, in their own time, extended and enriched the number and quality of contributions. The PSTs were able to utilize knowledge to consider their own teaching as compared to others which supported reflection. They gained new insights into ways to change practice.

As teacher education programs evolve into hybrid or totally online, it is important to maintain both the principles of effective online instruction and the steps of the lesson study cycle. As more variations will inevitably occur, to maintain the integrity of the original lesson study, as practiced effectively for more than a century, we need to be rigorous in describing and discussing the adaptations made for use with PSTs when online (Larssen et al., 2018). Technologies will continue to change and evolve. Video capture, coding of student behaviors, and coding of lesson study discussions are all new areas of research. Another challenge of online lesson study is the consideration of equity. This includes equity in using technology, an underresearched topic (Huang et al., 2021). While online lesson study may be a way to provide greater support, flexibility, and access for teachers, addressing the digital divide remains a challenge. To advance online lesson study, we must consider using increasingly inclusive, equitable, and compassionate ways to teach.

References

Creswell, J. W. (1998). *Qualitative inquiry and research design: Choosing among five traditions*. Sage Publications, Inc.

Huang, R., Helgevold, N., & Lang, J. (2021). Digital technologies, online learning and lesson study. *International Journal for Lesson and Learning Studies, 10*(2), 105–117. https://doi.org/10.1108/IJLLS-03-2021-0018

Larssen, D. L. S., Cajkler, W., Mosvold, R., Bjuland, R., Helgevold, N., Fauskanger, J., & Norton, J. (2018). A literature review of lesson study in initial teacher education: Perspectives about learning and observation. *International Journal for Lesson and Learning Studies, 7*(1), 8–22. https://doi.org/10.1108/IJLLS-06-2017-0030

National Research Council. (2013). *Next generation science standards: For states, by states*. The National Academies Press. https://doi.org/10.17226/18290.

Ward, J. R., & McCotter, S. S. (2004). Reflection as a visible outcome for preservice teachers. *Teacher and Teacher Education, 20*(3), 243–257.

Weaver, M. G., Goedde, A. M., Nadler, J. R., & Patterson, N. (2021). Digital tools to promote remote lesson study. *International Journal for Lesson and Learning Studies, 10*(2), 187–201. https://doi.org/10.1108/IJLLS-09-2020-0072

Windschitl, M., Thompson, J., & Braaten, M. (2020). *Ambitious science teaching*. Harvard Education Press.

Yin, J., & Partelow, L. (2020). *An overview of the alternative teacher certification sector outside of higher education*. Center for American Progress. https://files.eric.ed.gov/fulltext/ED615834.pdf

14

USE OF STRUCTURED VIDEO REFLECTION TO DEVELOP TEACHER NOTICING DURING LESSON STUDY

Gillian Roehrig and Jennifer Suh

This chapter focuses on structured video analysis within the *reflect* step of lesson study. A major challenge within the *reflect* step is making sure that the discussion enables learning. Reasons for failure to enable learning are "typically related to the quality of data collected or the quality of the process used to present and discuss it" (Lewis et al., 2019, p. 31). Lewis et al. (2004) described a range of data collection approaches, most common being the collection of detailed narrative records. While not a requirement of the *reflect* step, many researchers use video to avoid problems in the *reflect* step related to the quality of the collected data (e.g., Huang et al., 2019). In this chapter, we present two cases that illustrate ways to support the success of the *reflect* step by using video to facilitate reflection supported by the Noticing Framework (van Es & Sherin, 2002).

Use of Video

Teacher educators have used video as a learning tool for several decades as it affords the ability to capture the richness and complexity of classrooms and allows teachers to examine and reflect on their teaching practices (Grossman, 2005). Video allows teachers to remove themselves from the demands of the classroom and to step back and examine classroom events (van Es & Sherin, 2008). With technological advances, for example the development of video annotation tools (McFadden et al., 2014), video has become a powerful tool to enable reflective practice (Rich & Hannafin, 2009).

Sherin (2003) described two affordances of using video in teacher education: "(a) video allows for a permanent record of classroom occurrences that can be viewed repeatedly to ensure capture of classroom complexity and student-teacher interactions, and (b) video provides the opportunity for teachers to develop an

DOI: 10.4324/9781003326434-19

'analytic mind set'" (p. 13). van Es and Sherin (2002) argue that teachers need to develop reflective and analytical skills to learn to notice from analysis of their own teaching. They found that teachers who used video annotation showed more improved reflective abilities than those in the control group. Subsequently, Sherin and van Es (2009) investigated the use of video annotation with mathematics teachers and reported on changes in teachers' ability to interpret student thinking.

Technological developments have transformed the ways in which video and video annotation tools can afford learning in teacher education. However, these technological advancements have necessitated the development of new frameworks to guide the utilization of video (Sherin, 2003). In our work, we draw explicitly on the Noticing Framework (van Es & Sherin, 2002, 2008).

Noticing Framework

van Es and Sherin (2002, 2008) proposed that the skill of noticing consists of two phases (describing and analyzing). The describing phase involves teachers being able to accurately describe a classroom event. However, not all events occurring within the lesson constituting the *teach* step, can or need to be attended to. Thus, the first characteristic of noticing is attending, "learning to identify what is noteworthy about a particular situation" (van Es & Sherin, 2002, p. 573). In the analysis phase, teachers are expected to move beyond a literal description of the event to thinking about why an event occurred. In taking an evaluating stance, teachers should determine what worked or could have been done differently, providing evidence to support their claims (van Es & Sherin, 2002). Finally, van Es and Sherin (2002) call for teachers to take an interpretive stance, drawing inferences and making "connections between specific events and broader principles of teaching and learning" (van Es & Sherin, 2008, p. 245). Specific to lesson study, the ability of teachers to successfully take an interpretative stance in the *reflect* step is impacted by the principles of teaching and learning attended to in the *study* step and the alignment of all steps of the lesson study cycle with these principles.

Noticing skills are described by Barnhart and van Es (2015) with levels of sophistication (Table 14.1). Ideally, as we work with teachers in lesson study, we want them to develop sophisticated noticing skills. Working with colleagues gives opportunities for teachers to develop this sophistication and to collectively make sense of teaching events. Finally, as the lesson study team thinks about future lessons, they have an opportunity to "respond" in ways where they identify and describe "acting on a specific student idea during the lesson and offering specific ideas of what do different next time in responding to evidence" (Barnhart & van Es, 2015, p. 88).

In summary, noticing refers to what teachers attend to in reflecting on their teaching, as well as how they reason about what they observe (van Es & Sherin, 2008). "It is critical for teachers to first notice what is significant in a classroom interaction, then interpret that event, and then use those interpretations to inform their pedagogical decision" (van Es & Sherin, 2008, p. 247). This view of noticing

TABLE 14.1 Levels of Sophistication for Noticing Skills (Barnhart & Van Es, 2015).

Skills	Low sophistication	Medium sophistication	High sophistication
Attending	Highlights classroom events, teacher pedagogy, student behaviors, and/ or classroom climate. No attention to student thinking	Highlights student thinking with respect to the collection of data from a scientific inquiry	Highlights student thinking with respect to the collection, analysis and interpretation of data from a scientific inquiry
Analyzing	Little or no sense-making of highlighted events; mostly descriptions. No elaboration or analysis of interactions and classroom events; little or no use of evidence to support claims	Begins to make sense of highlighted events. Some use of evidence to support claims	Consistently makes sense of highlighted events. Consistent use of evidence to support claims
Responding	Does not identify or describe acting on specific student ideas as topics of discussion; offers disconnected or vague ideas of what to do differently next time	Identifies and describes acting on specific student ideas during the lesson. Offers ideas about what to do differently next time	Identifies and describes acting on a specific student idea during the lesson and offers specific ideas of what to do differently next time in response to evidence; make logical connections between teaching and learning

is aligned with knowledge integration theories that provide theoretical framing for lesson study.

Case Studies

In the following section, we provide two case studies to illustrate the use of noticing through video in conjunction with the *reflect* step of lesson study. Given the affordances of video, both cases used video for the structured *reflect* step. Lewis et al. (2019) recommend the use of a protocol as a shared repertoire to guide the post-lesson discussion,

> Using a discussion protocol increases the likelihood that the post-lesson discussion will focus on presentation and discussion of observers' data, with a focus on the ideas posed by the team.

(p. 32)

Similarly, we focused on having a structured lesson analysis tool to help guide discussion during the *reflect* step. These cases illustrate how a structured reflective tool, used in conjunction with video, during the *reflect* step helped teachers to explicitly focus their lesson documentation and reflection on the goals established in the *study* and *plan* steps. The first case explores a preservice integrated Mathematics and Science Methods course that used lesson study for a STEM problem-based learning (PBL) unit. The second case explores the use of a modified lesson study within a professional development for in-service elementary science teachers. The presentation of both cases is guided by the following questions: In what way did the structured video reflection coupled with LS reflection (a) allow participants to affirm, question, and consolidate ideas about teaching and learning about mathematics and science; (b) what ideas were reflected on and refined through teacher noticing?

Case Study 1

In Case 1, elementary preservice teachers (PSTs) co-designed and co-taught a STEM PBL unit. After the lesson implementation, they individually annotated the video recorded lesson followed by a collective debrief of the lesson activities that provided diverse learners access to the content and engagement in inquiry-based learning. The instructor modified the Culturally Responsive Mathematics Teaching (CRMT) Lesson Analysis Tool (Aguirre & Zavala, 2013) designed to promote intentional teaching discussions and critical reflection on integrated mathematics and science lessons. The CRMT was specifically designed as a self-reflective tool to support lesson/unit design and implementation, which was ideal in the lesson study setting to provide alignment across the steps of the lesson study cycle. Findings revealed that PSTs effectively used the CRMT lesson analysis tool when reflecting on their lessons, focusing on equitable teaching practices. They made meaningful connections across the CRMT categories with their lesson enactment by annotating instances where they felt that they attended to practice or felt there was a missed opportunity. This focused individual attention allowed PSTs to collectively analyze and consider how they may respond in the future. This time for professional noticing enhanced their reflective practice in delivering equitable instruction for all their diverse learners and developed their vision for assessing strength in children.

As an example, one team of four PSTs, (Lea, Helen, Joanne, and Jenna) planned a STEM lesson where students used mathematics to make decisions, weighing the cost and benefit of creating a new area of the national park instead of leasing this land to an oil company. The *reflect* step focused on a specific mathematics lesson where students were creating the shape of the park and finding the area. After the lesson, the team was asked to watch and annotate the video using the CRMT lesson analysis tool. We share a few of the PSTs' video-based reflections to illustrate how PSTs affirmed, questioned, and consolidated ideas about teaching and learning about mathematics and science. For example, Lea annotated early in the lesson to reflect on how she anticipated and supported student thinking,

> When we were planning the unit, we mistakenly anticipated that the students would already know how to create the irregular polygons. I know for myself, I did not conceive of how I might adequately support groups in creating their shapes, as I thought they would already know this content. I should have considered how to support them so that they did not need to count 300–500 squares.

In this annotation, she affirmed the high level of cognitive demand of the task of figuring out the area of irregular polygons and questioned how she could have better supported the students. She not only identified and described not being ready to respond to student ideas during the lesson but also reflected on the idea that she needs to do something different next time stating, "I should have considered how to support them so that they did not need to count 300–500 square." She talked with her team and the instructor about how to decompose irregular shapes to determine areas and combine fraction parts to get to a reasonable number of square units.

In another example, Lea complimented Helen for a teacher move that focused on distributing intellectual authority during group work indicating how the *reflect* step allowed her to recognize the practice of a colleague which helps her improve upon her knowledge,

> I think Helen did a great job of prompting the group to collaborate here. This group in particular had a large personality that tended to dominate, precluding the participation of other group members. However, the best learning occurs in collaboration. By encouraging the group to work together, Hannah gave the less dominant group members the opportunity to learn more about area and perimeter than they would have otherwise.

This annotation shows how the PSTs were consolidating a big idea around *power and participation* and recognizing it when they saw a colleague implementing a strategy to center equitable teaching practices. Lea not only attended to and analyzed Hannah's teacher moves but responded to this teaching practice to distribute intellectual authority.

Later in the lesson, Jenna annotated on the ways she *celebrated students' thinking* and the ways she supported deepening their understanding of the area of a circle.

> This group really wanted to draw a perfect circle for their perimeter. They had done several calculations to figure out the radius and diameter and I brought over the string to show them a way of using that information to draw the circle. I think this was a good effort to help them move past a stuck point, but I think they still needed more explicit instructions on HOW to do this.

Here she not only shows evidence of what she attended to in student thinking but also demonstrates how she analyzed in the moment and decided to respond to support student thinking.

This case illustrates how PSTs used video and a structured observation tool to distinguish and consolidate new ideas around teaching through culturally relevant pedagogy. This structured reflective process using the video annotation tool provided opportunities to make sense of which events to attend to, as well as analyze using evidence to make claims and mark specific students' ideas during the lesson and opportunities for future modifications based on the evidence to better support equitable mathematics teaching.

Case Study 2

This case is drawn from a teacher professional development program designed to promote reform-based science teaching practices, Science Teachers Learning through Lesson Analysis (STeLLA). STeLLA is aligned with features of quality professional development (e.g., Darling-Hammond et al., 2017) and has been shown to have a significant impact on elementary teachers' science content knowledge, pedagogical content knowledge, and instructional practices, as well as student achievement in science as compared to a control group (Roth et al., 2011; Taylor et al., 2017). STeLLA is a year-long professional development program, beginning with an intensive 2-week summer institute (study and plan) with academic year follow-up through a series of small study groups of the professional development participants (teach and reflect). Teacher learning is guided by a conceptual framework that supports teachers in making student thinking visible and the development of a coherent science content storyline (Roth et al., 2017). During the summer, teachers engaged in content deepening and analysis of practice activities supported through video cases of the STeLLA strategies and educative classroom science curricula. These curricula were designed to model and scaffold the teachers' use of the strategies and support teachers in deepening their science content knowledge. In the academic year, the teachers taught using the STeLLA curriculum and met monthly with their study groups to analyze video clips from one another's implementation of the curriculum.

Central to the success of STeLLA is lesson analysis through video, guided by the STeLLA conceptual framework. Teachers are first introduced to this lesson analysis process during the summer where they analyze videos of experienced teachers implementing STeLLA curricula. This allows teachers to become familiar with the analysis process and deepen their ability to notice specific STeLLA strategies, as well as improving their ability to analyze and respond to student thinking. This supports the teachers' ability to successfully engage in the reflection on their own teaching of the STeLLA curriculum. Lesson analysis focuses on short video clips from a classroom where the curriculum is being implemented, each clip focuses on a specific STeLLA strategy (Roth et al., 2017). Teachers watch the video multiple times to generate a claim with supporting evidence and reasoning related to the target STeLLA strategy. Teachers are also encouraged to propose alternative teacher moves to promote student learning (Taylor et al., 2017).

The *reflect* step of the modified lesson study cycle engaged the teachers in video analysis using short video clips from their implementation of the STeLLA curriculum selected by the professional development facilitators and a structured analysis using the STeLLA strategies to guide noticing. The video clips were purposefully chosen to focus on a small number of specific STeLLA strategies. In addition to the video clip, a transcript was provided which the teachers were expected to use to provide evidence for any assertions or claims they made related to the use of the STeLLA strategies, supporting higher level noticing skills of analyzing and responding. In addition, teachers were expected to draw on the STeLLA strategies booklet to ground their conversation. In the following example, the teachers are focusing on how to use elicit, probe, and challenge questions to support student learning. The facilitator started the video analysis by reminding the teachers about the purpose and to promote the integration knowledge of STeLLA strategies into their reflections,

> We all know when you're in the classroom and a kid says some off the wall idea, which they always will. At that point, we have to make that split second decision. What do I question? What do I not question? .... At full speed, these decisions are all a lot harder and a lot messier. That's a lot easier to do at the speed of video analysis. So, we have the transcript in front of us. We can take the time in this space to say, "When that student said this, I wonder... so this is meant to be a space, where we can practice that process to give ourselves a little bit of a leg up at full speed.

The teachers individually identified instances of elicit, probe, and challenge questions, as well as looking for missed opportunities. The facilitator then led a collaborative reflection on the use of elicit, probe, and challenge questions. For example,

Teacher 1:   I think it's a challenging question at 1:25, at the bottom of that page.
Facilitator: And why does that feel like a challenge question?
Teacher 1:   It's kind of gone over that same idea with the probing that the teacher was doing, but then she specifically tries to tie it to something and said, what happens?
Teacher 2:   I had also written that down .... because I was thinking the kids are... So I don't know exactly what the kid had been showing her, but it sounds like the kid had been showing one type of movement with foam mats. And then she said... she changed the idea. She was like, so what would happen if the arrows moved this way instead? Which is why I put it as a challenge question.

Following a discussion about another example of a challenge question, the teachers discussed missed opportunities. As the discussion veered away from a focus on the STeLLA strategies, the facilitator re-focused the conversation on the STeLLA strategies and supported their knowledge integration by summarizing two different ways to think about challenge questions in the STeLLA strategies booklet.

If we think back to the [strategies booklet], one way or one version of a challenge question is an introduction of new vocabulary or a new idea. So a challenge question doesn't always have to be like pushing against an idea that a student presented, it can also be the teacher, as you all had noticed at 1:22, right? …When you all explained it, if she moved the arrows around and she was introducing a new idea, then that could have been a challenge question because she was introducing this new situation for them to analyze.

The facilitator wrapped up the study group by asking the teachers to reflect on their process during the study group and why it was important to think about and distinguish between these types of questions. Teacher 3 responded,

It makes me think that I really need to make sure that my content knowledge is up to speed so that when I ask those questions, I know what to do with their answers. I know how to utilize that information.

Like Lewis et al. (2019), we argue for the role of a knowledgeable other to facilitate aspects of the lesson study cycle. The experiences of the STeLLA professional development further this line of reasoning, arguing that this external facilitation should be driven by a clear theory of action and specific framework to support teaching noticing. Using the framework to guide the *reflect* step maximizes teachers' learning related to specific professional development learning goals. Roth et al. (2011) compared two professional development approaches both of which used video analysis but only one group using a framework-driven approach to video analysis (i.e., STeLLA). In the other group, the teachers determined which aspect of the video was of interest to them and were allowed to develop self-directed inquiries into their practice. While this group had improved attitudes toward teaching science and improved self-efficacy in driving their own professional development, their inquiries lacked the focus to have any significant impact on teaching practice. In contrast, when reflection was guided by the STeLLA framework, teachers were able to implement specific changes in their teaching.

Concluding Thoughts

The Lesson Study *reflect* step can take many forms, however as we prefaced in the description of the cases, we chose to use video to support teacher reflection. Both cases illustrate the critical role of structured reflective tool (i.e., CRMT and

STeLLA strategies) during the *reflect* step, and the use of video allowed us to slow down practice and promote more sophisticated levels of noticing (van Es & Sherin, 2008). By using video, teachers could watch an event multiple times to support deeper understanding of student thinking and teacher practices (such as elicit, probe, and challenge questions). Rather than moving directly from the *teach* step into a collective *reflect* step, both cases first provided an opportunity for individual reflection followed by collaborative meaning-making and reflection grounded in evidence from the video and the guiding framework. The opportunity to build on each other's ideas and discuss multiple perspectives provides a rich environment for sophisticated noticing and reflection. The first case utilized a video of an entire lesson given the context of a university methods course, whereas the second case used short video clips given the time constraints of a professional development setting. When carefully selected by an expert facilitator, short video clips are still able to elicit sophisticated levels of noticing. This allows a teacher educator to maximize the often-short time available in a methods course to focus reflection on specific strategies or issues and address specific learning goals established within the *study* and *plan* steps.

The two cases exemplify how two educators used lesson study with structured video reflection to help teachers distinguish and consolidate ideas and move teachers to more sophisticated levels of noticing. First, the specific frameworks used in each case provided teachers with common language and nameable strategies to enhance their practices. For example, when PSTs reflected on cognitive demand, depth of knowledge, power, and participation, and cultural funds of knowledge, they all understood these criteria as the CRMT framework was used across all steps of the lesson cycle. Similarly, in STeLLA, teachers distinguished from practices that "probe student ideas and predictions" versus "engage students in analyzing and interpreting data and observation." The focus around these practices highlighted not only classroom events but allowed teachers to highlight student thinking that was elicited with respect to the classroom discourse and the exchanges of ideas. This high level of attention to student thinking was supported through the use of the video as a "permanent record" of practice to move beyond a literal description of the event to considering different approaches to teaching and learning and reflect on and refine on the effective teaching practices. At the same time, if the event did not go as anticipated, it also allowed for teachers to reflect on alternative strategies for the future and reflect on how to respond in those situations, which allows for the consolidation of their professional learning. This framework- and video-supported approach to the *reflect* step of the Lesson Study cycle supports knowledge integration by providing an "in-depth opportunity for the last *two stages of knowledge integration: distinguishing and consolidating ideas and ideas are reflected on and refined so that they fit together"* (Lewis et al., 2019, p. 30).

Finally, we underscore that the success of the *reflect* step is only as good as the preceding *study*, *plan*, and *teach* steps. In both cases, the *study* step was supported by strong frameworks for reform-based teaching that teachers were supported in learning through either the methods course or professional development program. Most critical, these frameworks introduced in the *study* step were threaded throughout the lesson study cycle, being explicitly used to design or modify units during the *plan* step and to guide reflection during the *reflect* step. Additionally, the *plan* step was supported by the use of quality curriculum, in the first case, this was developed with the support of the methods instructor and in the second case, the teachers were provided with educative curricula to use in their classrooms. Educative curriculum has been shown to support the development of elementary teachers' reform-based science pedagogies (e.g., Arias et al., 2016). By providing curricular materials, PSTs and in-service teachers are better supported in implementing the desired pedagogies during the *teach* step, leading to a *reflect* step that supports reflection on the desired pedagogical approaches. Finally, as noted by Lewis et al. (2019), if data collection is not aligned with the learning goals of the lesson and focused on student thinking, the *reflect* step will be compromised. The use of video provides a strong record of instruction and student thinking that forms the basis for a successful *reflect* step.

References

Aguirre, J. M., & Zavala, M. (2013). Making culturally responsive mathematics teaching explicit: A lesson analysis tool. *Pedagogies: An International Journal*, *8*(2), 163–190.

Arias, A. M., Davis, E. A., Marino, J., Kademian, S. M., & Palincsar, A. S. (2016). Teachers' use of educative curriculum materials to engage students in science practices. *International Journal of Science Education*, *38*(9), 1504–1526.

Barnhart, T., & van Es, E. (2015). Studying teacher noticing: Examining the relationship among pre-service science teachers' ability to attend, analyze and respond to student thinking. *Teaching and Teacher Education*, *45*(4), 83–93.

Darling-Hammond, L., Hyler, M. E., & Gardner, M. (2017). *Effective teacher professional development*. Learning Policy Institute.

Grossman, P. (2005). Research on pedagogical approaches. In M. Cochran-Smith, & K. M. Zeichner (Eds.), *Studying teacher education* (pp. 425–476). Lawrence Erlbaum.

Huang, R., Kimmins, D., & Winters, J. (2019). A critical mechanism for improving teaching and promoting teacher learning during Chinese lesson study: An analysis of the dynamics between enactment and reflection. In *Theory and practice of lesson study in mathematics* (pp. 705–730). Springer.

Lewis, C., Friedkin, S., Emerson, K., Henn, L., & Goldsmith, L. (2019). How does lesson study work? Toward a theory of lesson study process and impact. In Huang et al. (eds.) *Theory and practice of lesson study in mathematics* (pp. 13–37). Springer.

Lewis, C., Perry, R., & Hurd, J. (2004). A deeper look at lesson study. *Educational Leadership*, *61*(5), 18–22.

McFadden, J., Ellis, J., Anwar, T., & Roehrig, G. (2014). Beginning science teachers' use of a digital video annotation tool to promote reflective practices. *Journal of Science Education and Technology*, *23*(2), 458–470.

Rich, P., & Hannafin, M. (2009). Video annotation tools technologies to scaffold, structure, and transform teacher reflection. *Journal of Teacher Education*, *60*(1), 52–67.

Roth, K. J., Garnier, H. E., Chen, C., Lemmens, M., Schwille, K., & Wickler, N. I. Z. (2011). Video-based lesson analysis: Effective science professional development for teacher and student learning. *Journal of Research in Science Teaching*, *48*(2), 117–148.

Roth, K.J., Bintz, J., Wickler, N.I.Z., Hvidsten, C., Taylor, J., Beardsley, P.M., Caine, A., & Wilson, C.D. (2017). Design principles for effective video-based professional development. *International Journal of STEM Education*, *4*(31), 1–24. https://doi.org/10.1186/s40594-017-0091-2

Sherin, M. (2003). New perspectives on the role of video in teacher education. In J. Brophy (Ed.), *Using video in teacher education* (pp. 1–27). Elsevier Science.

Sherin, M., & van Es, E. (2009). Effects of video club participation on teachers' professional vision. *Journal of Teacher Education*, *60*(1), 20–37.

Taylor, J. A., Roth, K., Wilson, C. D., Stuhlsatz, M. A., & Tipton, E. (2017). The effect of an analysis-of-practice, videocase-based, teacher professional development program on elementary students' science achievement. *Journal of Research on Educational Effectiveness*, *10*(2), 241–271.

van Es, E., & Sherin, M. G. (2002). Learning to notice: Scaffolding new teachers' interpretations of classroom interactions. *Journal of Technology and Teacher Education*, *10*(4), 571–596.

van Es, E., & Sherin, M. (2008). Mathematics teachers' "learning to notice" in the context of a video club. *Teaching and Teacher Education*, *24*(2), 244–276.

15

PROMOTING INTERDISCIPLINARY CONNECTIONS AND MATHEMATICS COLLABORATIONS THROUGH LITERACY-CENTERED LESSON STUDY

Hanna Haydar, Meral Kaya, and Joanna Weaver

Introduction

Interdisciplinary collaboration through lesson study provides opportunities for teachers to share comprehensive and effective instructional ideas through research, study, and reflection. Researchers have explored how teaching both generalizable and disciplinary literacy supports more comprehensive and connected learning of content and skills (Dobbs et al., 2016). Interdisciplinary collaboration emerged to connect various curriculum areas with literacy leading to more effective learning (Fang, 2012). Interdisciplinary collaboration facilitates, for example, mathematics content learning and develops literacy skills.

Integration of literacy with elementary mathematics requires teachers to combine both their mathematics and literacy expertise into the process of planning successful interdisciplinary lessons. Lesson study can establish a platform for preservice teachers (PSTs) to collaborate in planning effective interdisciplinary lessons, yet limited studies exist that examine what elementary PSTs from mathematics and literacy methods classes perceive and experience when participating in the lesson study process.

This qualitative study exemplifies how two elementary methods courses integrated literacy into mathematics instruction while modifying lesson study groups to facilitate PSTs' use of children's literature and storytelling when teaching elementary mathematics. The first two authors were the instructors of the two methods courses and implemented the integration of literacy and mathematics, and the third author helped in the analysis of the data and chapter writing. In this chapter, we will examine the impact of integrating lesson study across mathematics and literacy methods courses as we unpack the context of the lesson study and the PSTs' interaction with the process.

DOI: 10.4324/9781003326434-20

Theoretical Background and Literature Review

Lesson study research teams are sometimes called "communities of practice" (CoP: Soto et al., 2019), and they focus on a community of educators who are sharing instructional practices and learning together. These communities may be content-specific or interdisciplinary (Weaver et al., 2021). Our research team was interdisciplinary and shared instructional practices across a literacy course that focused on children's literature and a graduate childhood mathematics education course.

We wanted to examine how lesson study affected teacher candidates' knowledge and skills of lesson planning. Lesson study helps develop PSTs' professional knowledge and skills while sharing experiences (Aykan & Dursun, 2021). Lesson study teams experience a deeper connection with the content, collaborative discussions of teaching strategies with peers, meaningful and purposeful observations of teaching and learning, and increased confidence in lesson planning compared to those teacher candidates not engaged in lesson study (Chassels & Melville, 2009).

Literacy as a Socially Constructed Interaction

Interdisciplinary approaches to teaching have gained attention because they offer more comprehensive learning opportunities (Uy & Frank, 2004). For example, because literacy is a tool students use in all content areas to comprehend (Donoghue, 2008), students receive a deeper understanding of mathematics using comprehension strategies and reasoning while organizing thinking by talking through problems. Furthermore, when students write step-by-step strategies to solve the problem and share their ways in solving problems in the class and hear different strategies to solve these problems, student understanding increases. Integrating literacy and math fosters students' ability to use and develop their literacy skills and gain math content to examine mathematical topics and problem-solving practices. Teachers can integrate literacy into any content area to build students' learning.

Teacher educators utilize lesson study because it offers a mechanism for interdisciplinary planning. Lesson study creates a framework for interdisciplinary connections and syntheses of instructional methods. Building on Section 2 of this book, recall that the *study* step provides the framework to examine and initiate thinking, and the *plan* step will empower voices, ideas, and collaboration to include the PSTs' expertise, skills, and knowledge in bringing mathematics and literacy together successfully. Lesson study will facilitate these skills and knowledge and create a guide and framework in their brainstorming, planning, and reflection (Lewis et al., 2019). Using lesson study with PSTs provides an opportunity to explore constructive and effective methods of instruction in the research lesson.

In mathematics instruction, children are expected to demonstrate various mathematical practices to conceptualize mathematics and/or solve a mathematics problem. An interdisciplinary approach strengthens this scaffolding through literacy skills and mathematics, and children's literature may play as the discourse and context. McDuffy and Young (2003) indicated that children's books act as context for mathematical thinking and problem-solving. Children's books can be powerful mentor texts and great models that can facilitate mathematics learning, serving as a conduit for using mathematical language, engaging in solving mathematical problems in a more authentic and real-life inspired way. When mathematics is embedded in a story, students see mathematics concepts more holistically. While students discuss and analyze mathematics concepts or create or solve mathematics problems through stories, teachers can design activities that develop and integrate both mathematics and literacy skills.

As stories create authentic contexts for mathematics problems, learning is significant and lasting (Durmaz & Miçooğulları, 2021). In lesson study, mathematics content embedded in children's literature leads to strengthened literacy skills and mathematical practices through collaboration. This modified lesson study explores and reflects how an interdisciplinary approach facilitated and promoted mathematical skills and content. Furthermore, it examines how children's literature empowered construction of mathematics and literacy learning.

Literacy-centered Lesson Study

Context and Research Goal

The first two authors implemented this study in their respective classes with two different groups of PSTs. The mathematics educator worked with 22 graduate-level PSTs in the context of a mathematics methods class for career changers, and the literacy educator worked with 19 undergraduate PSTs in a course called *Integrated methodology for literacy instruction*. The lesson study spanned 12 sessions in both classes and followed the same structure starting with two sessions for the *prepare* step, which included a lesson study orientation, followed by three sessions for *study* and four for *plan*. The *teach* step was replaced with teaching rehearsals due to the COVID-19 pandemic and two sessions were given at the end for groups to *reflect* and present their findings to the rest of their classes.

Both researchers served as knowledgeable others who provided final comments during the reflection (Takahashi, 2014) in each other's classrooms to give insights and feedback about the other content area. The literacy educator presented to students in the mathematics course about the characteristics of children's literature in interdisciplinary teaching, modeled examples of teaching through stories focusing on literacy skills and mathematical concepts and practices, and provided resources on literacy teaching and teaching through children's literature. The mathematics educator visited the literacy class and presented about using stories as contexts for mathematizing, modeled examples of how stories can be used to generalize mathematical models or guide multi-step problem solving, and provided resources on elementary mathematical standards. All materials were posted on the courses' learning management system and instructors remained available through scheduled content conferencing. To proactively anticipate and address the possible content gaps, the researchers developed teacher guides to help PSTs put student thinking and learning at the center of the lesson planning activity when using children's stories to teach mathematics (Kaya & Haydar, 2020).

Our main goal in this collaboration was to support PSTs in integrating children's literature and mathematics and planning interdisciplinary lessons to support mathematizing.

Data Collection

In both classes, this study focused on the *study* and *plan* steps of the lesson study cycle and used rehearsals (*teach*) to enact parts of the teaching activities and concluded with a modified *reflect* step. Group meetings and rehearsals were videotaped or audio recorded and the collaboration reports and unit plans were submitted using the Teaching-Learning Plan Template (http://lessonresearch.net).

All tools, templates, and technological collaboration outcomes were copied and/ or collected. The researchers used a pre-assessment to capture PSTs' knowledge, experience, and self-efficacy beliefs about using children's literature/storytelling in teaching mathematics. While some of the questions focused on the PSTs' basic knowledge and experience in teaching with children's literature in general and teaching mathematics more specifically, the self-efficacy part of the survey focused on their belief in their own ability to succeed with this type of teaching. A post-assessment survey reflection was conducted at the end of the lesson study to attend to the participants' learning.

Data Analysis

Video and audio data were analyzed by determining critical events (Powell et al., 2003) that revealed how the modified lesson study facilitated PSTs' use of children's literature/storytelling to teach elementary mathematics. These events were then transcribed. Transcriptions and reflection journals were coded. In line with methods from grounded theory, the transcripts and journal data were sorted, re-sorted, and analyzed, moving from descriptive information to constructing explanatory schemes (Corbin & Strauss, 2014).

Anticipated Challenges and Lesson Study Modifications

When planning for this study, we anticipated challenges based on experience and the lesson study literature (Lewis et al., 2019), which helped introduce related modifications to the lesson study steps. For example, researchers anticipated that PSTs would emphasize product rather than process. Students develop this narrow focus on outcomes over their schooling years; therefore, it can be challenging to deconstruct and create a mindset of looking at the holistic picture instead. The instructors aimed to strengthen students' understanding of the relationship between the individual lesson and broader learning experience. In the *prepare* step, they designed introductory activities that focused on group norms and roles and built-in assessment tasks that monitored the submission of group meeting minutes to ensure that lesson study sessions and protocols were followed with fidelity. In the *study* step, they also provided resources for curriculum coherence maps that situate individual standards within the larger curricular framework.

Also, researchers anticipated gaps in PSTs' content knowledge, especially because the undergraduate participants did not take any mathematics class prior to this study. Before they create their mathematics units around children's literature, PSTs should be able to implement mathematics and literacy content knowledge and skills in collaborative unit plans. To address this, the researchers acted as content experts in each other's classes as mentioned above during the *prepare* step, so they covered both the literacy and mathematical needs of their students. They provided content presentations and curricular resources and held content-based office hours. Then during the *study* step, PSTs could focus and study both sets of resources across content areas as they pulled their ideas together.

PSTs may find it challenging to focus primarily on student learning when planning research mathematics lessons around children's literature. Bringing literacy and math together might be challenging so researchers modified the lesson plan template to include themes centered around this integration and required each lesson to include both mathematics and literacy standards and teaching objectives.

Findings

Lesson Study as a Vehicle to Learn to Integrate Mathematics and Literacy

During the *study* step, PSTs developed research themes related to teaching mathematics through children's literature and decided on possible mathematical and literacy topics that could be taught through the selected children's book and aligned to standards and curricular materials. In the *plan* step, PSTs situated the research lesson within the context of the overall unit and determined how the same children's book could be a context for a thematic unit covering multiple mathematics and literacy standards. Participants then planned the lesson and anticipated students' thinking and responses on each of the story-based mathematical lesson tasks. For the last step, participants practiced teaching parts of the activities they planned with each other as the audience, reflected about the whole process, and presented to the whole class about their learning and findings.

The analysis of the participants' post-reflections revealed that the main elements of the lesson study steps were effective contexts to learn about how to integrate literacy and mathematics. Three themes emerged that include: (1) Anticipation of student responses, (2) value of collaboration, and (3) perceptions of lesson study research lessons.

Anticipation of Student Responses

"Anticipating student thinking is a core teaching skill that allows teachers to notice student thinking and build bridges from it to new understandings" (http://lessonresearch.net). This element of lesson planning was introduced to participants through the lesson study template. In the *study* step, when PSTs studied standards, research, and curriculum materials related to their topics, the instructors prompted PSTs to review the course readings about the development of mathematical and literacy concepts to understand children's typical strategies and misconceptions. In the plan step, PSTs used this data when anticipating children's mathematical and literacy responses during the research lesson. This emphasis by the instructors and the lesson study template helped make this anticipation of students' responses significant and explicit.

Our analysis of the lesson study reports showed a pattern of participants anticipating children's mathematical misconceptions based on their research and suggesting the use of elements from the children's stories to address them. For example, a group that designed a lesson about counting by five in kindergarten anticipated that some students might still count by ones and suggested teachers should refer to the specific page from the story *Disappearing Moon* (Mind Research Institute, 2020) where it models using hands and tallies to count by five. This group's anticipation and decision-making at the plan step was facilitated by two elements from the *study* step: their investigating research and articles on development of number sense and the work they did with both instructors about the role of the story as a context for mathematizing.

Another group planned a third-grade research lesson about finding patterns when adding consecutive whole numbers. They anticipated some children might add the numbers randomly and suggested that teachers could ask students if they could use Cubey's method (Cubey is the main character in the story they chose for this unit). Again, this is indicative of what this group learned in the *study* step where they focused on how third graders can find and generate patterns and express regularity in repeated reasoning. The way they used the main character of the book to scaffold students' possible gap echoes what they learned about the use of children's literature in teaching.

Participants appreciated this aspect and considered it as a main element for teacher readiness, specifically stating, "anticipating students' responses can also prepare you so that you have an idea of what to expect and how to handle any situation that you are faced with and avoid being put on the spot." In addition, other PSTs said that children's thinking was at the heart of the planning stage. For example, one noted, "We have to plan in a way that accounts for all of our students, this would mean to think about what misunderstandings our students might have and resolve them beforehand."

Value of Collaboration

Collaboration is one of the major components of lesson study that teachers value the most (Lewis et al., 2004). Participants in this study noted in their final reflections how this collaboration became a meeting point for multiple perspectives: "I believe that teachers should collaborate because everyone brings their own unique perspective to the table." That came with its challenges of course: "Working with colleagues can be difficult and hard for everyone to stay on task." Participants stressed communication and flexibility as main facilitators of successful collaboration: "It's important to listen to each other's ideas and keep an open mind." This conversational collaboration affects the quality of the lesson study outcomes:

> I think because of our communication, we were able to create a lesson that flows nicely. If we just split the work up without talking it out with each other, we might have ended up creating a completely different lesson that makes no sense.

Other participants indicated the significance of collaborating when interdisciplinary topics are involved since it requires a different set of skills and expertise: "I learned how to work with others with different expertise to complete a lesson/unit plan. This can be especially helpful when working on how to connect other subjects together like we did with mathematics and literacy." For undergrad preservice, this collaboration made the interdisciplinary process tangible:

> I also learned that there are so many ways to teach a lesson that combines literature Math, or Science. It seemed difficult at first but once I really talked about it and worked on it with my group mates, it became clearer and more valuable.

Few PSTs saw the lesson study collaboration as a rehearsal to the professional collaboration that will be required from them in schools. This was voiced from PSTs preparing to teach in inclusion co-teaching models:

> As I am preparing to be an Integrated Co-Teaching (ICT) educator, the most important takeaway for me was collaboration with peers. In teaching we will often have to collaborate with other professionals and need to take into consideration individual ideas on how to teach a lesson. This lesson study taught me that every member has valuable input that will benefit students' learning.

Perceptions of Lesson Study Research Lessons

According to Lewis et al. (2019), the *plan* step should be a vehicle to develop teachers' knowledge, beliefs, and professional community. Participants in this study mentioned aspects of their collaborative lesson planning among their main takeaways. They learned how the instructional unit and lesson plans should be aligned with the lesson study goal: "I realize that you have to plan and prepare the research lesson so that all the activities are connected to the goal of the study." They also learned to perceive lessons as part of a larger unit: "we should remember that we are not creating one lesson and then move on to another lesson but that teachers should plan for continuous lessons that go hand in hand." For many, this was their first time working on unit planning and looking at an individual lesson in the context of the whole unit. The collaboration with other team members and the usage of the planning template helped make things clearer and offered a step-by-step manageable process:

> At first when I looked at one unit plan sample, I thought it would be very hard. Now that we were able to work on it as a group, I felt more comfortable with it. This also gave me practice with writing prompting questions as well as coming up with anticipated responses.

One PST expressed explicitly how the unit planning constituted a vehicle for learning and collaboration: "I learned a lot from discussing with my peers and putting ideas together to form a Unit and lesson plans. I thought it was great to hear from different perspectives and think differently about activities that could be conducted."

Integration of Mathematics and Literacy

We examined what PSTs learned about using literacy as context for teaching mathematics through children's literature. The pre-assessment survey revealed that most participants did not have any experience using children's literature to integrate literacy and mathematics instruction. To establish a baseline about their familiarity with various literacy practices such as reading aloud, silent reading, and discussion, 75% of the graduates and 60% of undergraduates reported they were either familiar or very familiar, whereas only 19% of the graduates and 10% of undergraduates reported they knew how to prepare activities from children's stories to teach mathematics.

The PSTs' participation in the modified lesson study shifted their responses from a mix of answers with less agreement with the importance of integration of literacy and mathematics to a stronger agreement after the lesson study. After the lesson study, 60% reported they were confident they knew how to prepare such activities and 20% were developing this knowledge.

Each group developed research themes centered around the integration of literacy to teach mathematics through children's literature and focused on problem-solving, and themes varied from "being creative collaborating to investigate mathematics problems presented in children's literature" to "using realistic fun stories which can encourage children to solve real-life problems using math," or "using mathematics literature to mathematize while showing various representations of mathematical problems and explaining mathematical thinking in different ways." These themes guided PSTs' collaboration in the lesson study steps. To better understand this shift and what participants took away, we analyzed the post-lesson study surveys and reflections. This analysis revealed themes related to the integration of literacy and mathematics.

Stories as Context for Mathematizing

An important shift occurred related to the participants' perception of the role of children's stories in teaching mathematics. In their pre-assessment survey, more PSTs perceived children's stories as mere motivation tools; whereas after the lesson study, a majority (75%) tended to understand children's literature as a "context for mathematizing," aligning with the research of Fosnot and Dolk (2002). We surmise this shift took place because preservice candidates had not experienced using the children's stories as anything more than motivation tools prior to the interdisciplinary experience. Following the experience and interacting with the children's stories as a context for learning mathematics and connecting with mathematics, they saw a broader use of the stories—as a "context for mathematizing."

This shift can be traced to the discussion and readings that the PSTs did in the *study* step about the role of context and the teacher guides and presentations in which the instructors modeled how stories can be used as contexts for mathematizing.

Within this view, PSTs ensured the selected book was developmentally relevant and met the curriculum standards: "Using the mathematics standards with the book is a good way to create lesson plans aimed towards the students' interest and grade level." The book becomes a source for problem solving investigations:

> I learned how to pick the right books for the students and ask meaningful open-ended questions. I think this will help me in the future to choose age-appropriate books for my students and to engage them in the classroom by asking questions during the reading leading to mathematics investigations.

This focus on text selection was discussed throughout the lesson study activities, in the *study* step, instructors discussed the features of good children's books to be used in an interdisciplinary approach. PSTs also analyzed, modified, and used a mathematics storybooks checklist that covered dimensions related to developmental level, literary merit, cultural relevance, and mathematical content and practices. At the end of the lesson study, participants reflected on the different children's books that they compared and selected.

As a context for mathematizing, participants learned they can refer to the story to develop or clarify mathematical concepts and strategies: "It is useful to refer back to stories when mathematizing because children can use that story to see how mathematics is everywhere and how a strategy can be done in different ways." This view affects the disposition that looks at mathematics as sense-making and related to daily life. For example,

> When you use these stories you make it possible for students to think about mathematics in a different manner. They can begin looking for patterns, relationships and think more deeply about math. The idea of mathematizing goes beyond arithmetic and algebra. We should be teaching mathematics around a context that children can relate to.

Literacy and Interdisciplinarity

Findings revealed that participants shifted their stance on the relationship between literacy and interdisciplinarity. In the pre-assessment, many PSTs expressed the idea that teaching mathematics through children's literature was intriguing but voiced concerns that they had no idea how this could be done and how challenging it might be. The lesson study helped them experience using interdisciplinary units firsthand and see the pedagogical value. They learned how children's books could become the context for multiple lessons and disciplines: "I also learned how it is possible to think of so many different lessons and subjects just by using one book." This places an importance on the book selection process "to make sure we choose a book that can be used across the curriculum. It is very important to ensure students get a full understanding of the book because it can be used in math, ELA, even science and social studies."

Some participants were able to generalize this concept of interdisciplinarity beyond just the children's literature and mathematics:

> I learned that finding literature that pertains to what you are teaching is vital and so is connecting other subject content matters with mathematics lessons. In general, it does not matter what subject you are teaching, you should always try to engage and relate other outside studies and content in your lessons.

The reflection activities in both the *study* and *plan* steps and after the rehearsals in addition to the instructors' emphasis and modeling of the interdisciplinary approach created a safe collaborative environment that supported these generalizations and creative thinking about interdisciplinarity.

Concluding Thoughts

This study illuminates how PSTs viewed interdisciplinary instruction when connecting mathematics and literacy through children's literature in the context of lesson study collaborations. Through lesson study, PSTs developed their interdisciplinary expertise and skills to plan lessons while strengthening their own learning. Results confirmed that collaboration through lesson study is key in extending skills and facilitating PSTs' understanding around integrating mathematics and literacy. Furthermore, the anticipation of students' responses prepared PSTs for more effective planning and instruction.

This study revealed that PSTs had an effective and impactful experience integrating mathematics and literacy through children's literature to teach interdisciplinary content. After providing lesson study within PST education, we recommend teacher educators utilize the lesson study process to promote interdisciplinary connections and collaborations.

References

Aykan, A., & Dursun, F. (2021). Investigating lesson study model within the scope of professional development in terms of PSTs. *International Online Journal of Educational Sciences, 13*(5), 1388–1408.

Chassels, C., & Melville, W. (2009). Collaborative, reflective and iterative Japanese lesson study in an initial teacher education program: Benefits and challenges. *Canadian Journal of Education/Revue Canadienne De l'éducation, 32*(4), 734–763. http://www.jstor.org/stable/canajeducrevucan.32.4.734

Corbin, J., & Strauss, A. L. (2014). *Basics of qualitative research: Techniques and procedures for developing grounded theory* (4th ed.). Sage.

da Ponte, J. (Eds.), (2019). *Theory and practices of lesson study in mathematics, an international perspective*. Springer.

Dobbs, C. D., Ippolito, J., & Charner-Laird, M. (2016). Layering intermediate and disciplinary literacy work: Lessons learned from a secondary social studies teacher team. *Journal of Adolescent and Adult Literacy, 60*(2), 131–139.

Donoghue, M. R. (2008). *Language arts: Integrating skills for classroom teaching*. Sage Publications, Incorporated.

Durmaz, B., & Miçoogullari, S. (2021). The effect of the integrated mathematics lessons with children's literature on the fifth grade students' place value understanding. *Acta Didactica Napocensia, 14*(2), 244–256. 10.24193/adn.14.2.18.

Fang, Z. (2012). Approaches to developing content area literacies: A synthesis and critique. *Journal of Adolescent and Adult Literacy, 56*(2), 103–108.

Fosnot, C. T., & Dolk, M. (2002). *Young mathematicians at work: Constructing fractions, decimals, and percents*. Heinemann.

Kaya, M., & Haydar, H. N. (2020). Children's literature in the landscape of teaching elementary mathematics: Examples for grades k to 5. In B. Durmaz, & D. Can (Eds.), *Matematik Öğretimi ve Çocuk edebiyatı*. Vize Yayıncılık.

Lewis, C., Friedkin, S., Emerson, K., Henn, L., & Goldsmith, L. (2019). How does lesson study work? Toward a theory of lesson study process and impact. In A. Takahashi, R. Huang, & J. da Ponte. *Theory and practice of lesson study in mathematics* (pp. 13–37). Springer.

Lewis, C., Perry, R., & Hurd, J. (2004). A deeper look at lesson study. *Educational Leadership, 61*(5), 18–23.

Mind Research Institute. (2020). *Disappearing Moon*. Retrieved February 15, 2023 from https://www.mindresearch.org/mathminds/stories

McDuffy, A., & Young, M. R. (2003). Promoting mathematical discourse through children's literature. *Teaching Children Mathematics, 9*(7), 385–389.

Powell, A. B., Francisco, J. M., & Maher, C. (2003). An analytical model for studying the development of learners' mathematical ideas and reasoning using videotape data. *Journal of Mathematical Behavior, 22*(4), 405–435.

Soto, M., Gupta, D., Dick, L., & Appelgate, M. (2019). Bridging distances: Professional development for higher education faculty through technology facilitated lesson study. *Journal of University Teaching and Learning Practice, 16*(3), 1–19.

Takahashi, A. (2014). The role of the knowledgeable other in lesson study: Examining the final comments of experienced lesson study practitioners. *Mathematics Teacher Education and Development, 16*(1), 2–17.

Uy, F., & Frank, C. (2004). Integrating mathematics, writing, and literature. *Kappa, Delta, Pi Record, 40*(4), 180–182.

Weaver, J. C., Matney, G., Goedde, A. M., Nadler, J. R., & Patterson, N. (2021). Digital tools to promote remote lesson study. *International Journal of Learning and Lesson Study, 10*(2), 187–201. https://www.emerald.com/insight/2046-8253.htm

16

ADAPTING LESSON STUDY FOR PRESERVICE TEACHERS' INSTRUCTION AND LEARNING

Joanna Weaver and Gabriel Matney

In their work with preservice teachers (PSTs), teacher educators sometimes need to be creative to provide pedagogical practices that extend PSTs' understandings of content, develop a foundation of instructional knowledge, and promote reflection and collaboration. Lesson study can strengthen these elements of PSTs' pedagogical practice through the *study, plan, teach, and reflect* steps (Lewis & Hurd, 2011). Lesson study is carried out by teacher teams working collaboratively to examine instructional ideas and their impact on student learning while lessons unfold. Lesson study is conducted in real time within the classroom. For many teacher educators, including ourselves, PSTs do not have ready access to enact lesson study in authentic K-12 classroom environments prior to their methods course and internship. Given this situation, we have been incorporating lesson study throughout the entire PST program (Matney & Fox, 2022).

Although our students are in the field every year, we do not yet have enough schools conducting lesson study and open to level of collaboration lesson study requires. Hence, when we think programmatically about how we can provide PSTs opportunities to enact lesson study, they must (a) learn about lesson study as a collaborative professional routine and (b) learn through enacting the lesson study process within their programs. We wondered with Lesson Study for Mathematics and Science Teacher Educators Conference (LSMSTEC) participants, given that the PST context differs from in-service, what adaptations can we make to lesson study that better serve the needs of budding teaching professionals? We hope this chapter and other chapters in this book help teacher educators answer this question for their own contexts. To help with that goal, we share our adaptation of lesson study, which we call Jigsaw Lesson Study in a pre-methods context and what we have found to be valuable from this adaptation.

DOI: 10.4324/9781003326434-21

PSTs participating in Jigsaw Lesson Study learn to collaborate with one another toward a common goal. They follow the same process as lesson study: *study, plan, teach, and reflect*. Based on conversations about lesson study with PSTs with the LMSTEC participants, we added the step *prepare* before the *study* step. This will be explained later in the chapter. The Jigsaw Lesson Studies are conducted in the classroom on campus prior to entering their full-time field placements. Jigsaw Lesson Study's name was derived from the jigsaw cooperative learning strategy that enables students to specialize in one aspect of a topic and then hold the responsibility of sharing that knowledge with others (Gillies, 2007). In the case of Jigsaw Lesson Study, each PST team studies different content, and then the content is shared through the enactment of the lesson. Our PSTs teach their lessons to three different teams of PSTs to engage in three cycles of *study, plan, teach, and reflect* within a course. When PSTs engage in collaborative practices such as Jigsaw Lesson Study, they have opportunities to work with peers to extend understandings of content, overcome instructional challenges with the support of their teacher education instructor as the knowledgeable other, and reflect and revise teaching ideas, while simultaneously learning about the lesson study process and its professional value.

Lesson Study for Preservice Teachers

There is evidence from research that engaging PSTs in lesson study is professionally beneficial. For example, in lesson study, collaboration increases awareness of the importance of studying, planning, teaching, and reflecting on instruction and the needs of the learners (Cajkler & Wood, 2016). Furthermore, Shuilleabhain and Bjuland (2019) reported that PSTs demonstrated instructional autonomy when designing learning goals. In addition, according to Matney and Fox (2022), PSTs reported an increased focus on instruction to improve student learning during their reflections, and they stated that this is much easier when they reflect collaboratively with peers. During the steps of lesson study, collaboration contributes to the development of communities of practice that work together to solve the challenges being faced with student learning or instruction (Weaver et al., 2021). Through collaboration in lesson study, PSTs feel a sense of belonging that enhances instruction, efficacy, and confidence (Matney & Fox, 2022).

Ogegbo et al. (2019) reported strengthened professional knowledge that included instructional strategies and content as well as increased enthusiasm and confidence levels when educators engaged in lesson study. According to Cajkler and Wood (2016), another benefit to conducting lesson study in a preservice classroom is the improvement of their observation skills, preparing them to experiment with innovative instructional ideas, and persevere together as a community of learners.

Çevik and Müldür (2021) revealed lesson study enhanced PSTs' planning skills and the ability to be flexible when working with others and conducting the lesson. Participants felt supported by their team and felt more confident in teaching and classroom management. Similarly, Sumarno's (2019) findings revealed that lesson study had a positive impact on PSTs' self-efficacy, confidence in student engagement, classroom management, and instructional strategies.

Bjuland and Helgevold (2018) noted the influential role of a facilitator and knowledgeable other to question, challenge, and focus PSTs' reflective dialog regarding evaluation of student learning and the effectiveness of instruction. Lesson study enhances PSTs' development of content and pedagogical knowledge, increases their confidence, and strengthens reflective habits needed to continuously grow and create engaging lessons for all students. As we designed Jigsaw Lesson Study for our PSTs, we wondered how: (1) They might acquire a deeper understanding of their content knowledge, (2) grow as collaborative educators and learners, and (3) develop a heightened awareness of how students learn. In what follows, we share the description and findings of Jigsaw Lesson Study as an adaptation for the PST context.

Jigsaw Lesson Study

Jigsaw Lesson Study (Figure 16.1) was adapted from lesson study (Lewis & Hurd, 2011). We made the adaptations to support PSTs' learning through enacting the lesson study process on campus prior to entering their field placements where they could engage in additional lesson study with in-service teachers. We overview the iterative Jigsaw Lesson Study process and then elaborate through the lens of the five steps of lesson study discussed in this book: *prepare*, *study*, *plan*, *teach*, and *reflect*. Prior to engaging in Jigsaw Lesson Study, our PSTs learned about each of the key steps of lesson study and the reason teachers around the world enact lesson study. We have modules and readings to facilitate this pre-learning. Also, they learned about teacher noticing, how to observe, and ways of reflecting. As seen in Figure 16.1, PSTs were grouped into teams of three, who work together to *study* specific elements of content, standards, and related curriculum. Next, the team *plans* a lesson. Then, the team *teaches* the full lesson to a different team of PSTs, with one team member teaching and the other two recording observational data. Following the first teaching, the team *reflects* on the data that include feedback from the learners and observers of the lesson, and the team considers revisions to improve the instruction. The cycle of teaching, reflecting, and revising occurs two more times (three times total). Hence, each of the three members will teach one time and observe two times. Each time the lesson is taught, it is taught to a new group of PST peers.

Prepare

The *prepare* step is completed by the teacher educator before facilitating *study, plan, teach, reflect* to prepare the PSTs for every step of lesson study. As mentioned in Chapter 4 by Hummer and Lesseig, during *prepare*, the teacher educator

Jigsaw Lesson Study

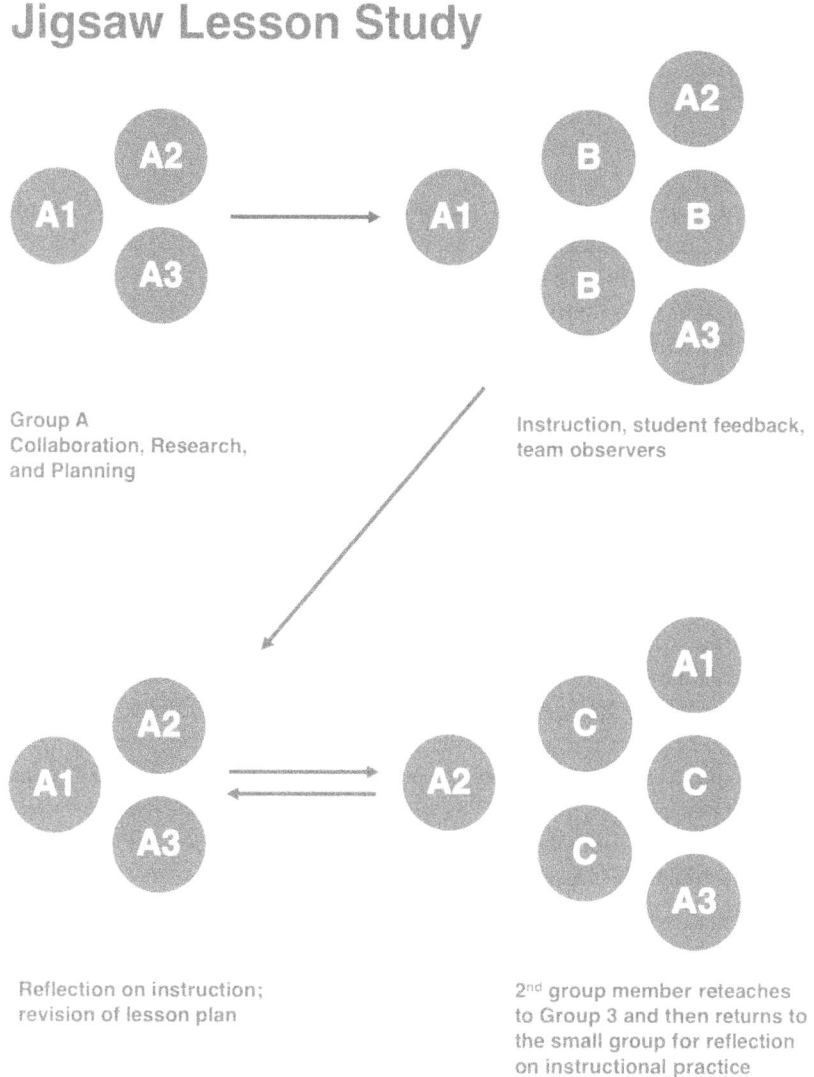

Group A
Collaboration, Research,
and Planning

Instruction, student feedback,
team observers

Reflection on instruction;
revision of lesson plan

2ⁿᵈ group member reteaches
to Group 3 and then returns to
the small group for reflection
on instructional practice

FIGURE 16.1 Jigsaw Lesson Study Flow. Weaver, G., & Fox, M. (2022). Jigsaw Lesson Study. BGSU, Bowling Green, Ohio.

introduces PSTs to lesson study, develops cooperative learning norms, and ensures proper access to instructional materials the PSTs will use throughout the process. Additionally, for Jigsaw Lesson Study, the teacher educator must prepare to be both the facilitator and at times, the knowledgeable other. In other words, when a teacher educator is enacting Jigsaw Lesson Study, they will play the role of knowledgeable other by providing the team with an outside voice on content, planning, reflection, and revision. Although it is possible to have a current teacher practitioner be a knowledgeable other, we have found this is not always possible due to school and university schedules and other constraints on teachers.

Study

During the *study* step, the PST teams brainstorm topics from the list of difficult-to-teach topics or from state data that reveal topics students find challenging. Teams must collaborate to come to the consensus on their chosen topics and then write their research lesson question. The PSTs are reminded by the teacher educator that the research question guides their study, planning, teaching, and reflection. Once the question is written, the teacher educator prompts the PSTs to explore various resources including, curricula, standards, teacher lessons, and articles about the topic. Throughout this process, the teacher educator acts as a knowledgeable other, asking probing questions about what should be studied and giving advice on where the PST might look to find helpful information on their particular topic of interest.

Plan

To begin the *plan* step, the PST teams share their findings they gathered during the *study* step. Teams are carefully instructed to move forward with their collaborative planning using the gathered data to inform their plan. In addition, they align the activities they are crafting within their plan in relation to their research question and anticipate both K-12 students and peer responses. Furthermore, PSTs consider how their instruction will strengthen student learning. Also, the team members decide the order they will teach and practice teaching their lesson to their small team before teaching other Jigsaw Lesson Study teams.

Teach

The *teach* step is organized carefully and explained to the PSTs. This step is where Jigsaw Lesson Study received its name since we have teams that studied different content, teaching one another throughout the lesson study process. Each team of three taught three times to three different teams (Figure 16.1); therefore, we form a larger group that includes four teams, and by the end of the three cycles, they will all teach one another. For example, Team A teaches Team B while Team C teaches Team D. After the first lesson is taught (i.e., Team A teaches Team B), the second team provides feedback and then teach their lesson (Team B teaches Team A) and receive feedback. When both teams have completed their teaching and received feedback, the individual teams meet to debrief, reflect, and revise their lessons, preparing for the second round of teaching of a new Jigsaw Lesson Study team. This concludes the first cycle of teach, reflect, and revise. This process is enacted within each cycle of the *teach* step. For the second round of teaching, Team A teaches Team C while Team B teaches Team D, and for the third round, Team A teaches Team D while Team B teaches Team C and vice-versa. A key component of the teach step is observation and data collection. We give PSTs a template for this and act as knowledgeable others during the lesson reflection, asking the PSTs about key noticings related to the research question and advising on future observations.

Reflect

Reflection is a key step throughout the Jigsaw Lesson Study process. We encouraged PSTs to critically think within the *study, plan, and teach* steps. The fourth step is *reflect*. We consider revision integral to the process for PSTs because it is their opportunity to fully engage in reflection using the feedback provided from other teams as well as the observation of their team members. They carefully consider the constructive feedback and observational notes to strengthen their lesson. The objective during the reflection time is to identify strategies that need clarification and to review student learning. Teams move through the various sections of their lessons and think about student learning and engagement. The revisions are deliberate as they continue to align and strengthen their lesson with their research question. They repeat this process after each set of lessons is taught, and then after their final lesson, they reflect on the entire Jigsaw Lesson Study process and how it impacted their professional growth.

Investigating Jigsaw Lesson Study

We found that Jigsaw Lesson Study is easily implemented in any content teacher education course such as mathematics, science, social studies, language arts, career tech workforce education, or Science Technology Engineering Art and Mathematics integration. For example, in a mathematics methods course, a Jigsaw Lesson Study team may choose a topic they believe will be challenging to teach or for students to learn such as *solving quadratic equations*. In a science or language arts methods course, teams may choose *using evidence to support a claim*. We first investigated Jigsaw Lesson Study in an interdisciplinary methods course of 55 PSTs, where teams focused on their own content specialization.

At the end of Jigsaw Lesson Study process, we collected data from survey responses that identified the PSTs' takeaways. We analyzed the data based on PSTs' perceptions of changes in instructional practice, professional knowledge, and their perception of the Jigsaw Lesson Study process. The constant comparative method (CCM) was utilized (Glaser & Strauss, 1967) with Erikson's (1986) coding process. Open coding provided for core categories to emerge as data were re-coded and reduced (Glaser, 1978; Glaser & Strauss, 1967; Strauss, 1987). We coded each set of surveys individually. After the second read-through, we discussed and negotiated assertions that emerged, and then we re-read the data with the agreed-upon assertions and met to clarify the examples that served as warrants.

Findings

Collaboration Influences the Jigsaw Lesson Study Topic and Instructional Practice

The survey responses revealed the PSTs' emphasis on collaboration, increased discussion, and how collaboration allowed Jigsaw Lesson Study teams to build off each team members' ideas and experiences. PSTs stated that innovative ideas were developed during discussions that expanded past experiences and helped them consider each team member's strengths. This created an overall flow of ideas and increased PSTs' confidence. Collaboration strengthened PSTs' understanding of the topic and contributed to the choice of the topic of focus of the lesson. PSTs felt supported by their team members and felt free to voice their areas of struggle. Additionally, PSTs identified common topics and viewed the lesson study experience as an opportunity to expand content knowledge.

Furthermore, findings revealed that collaboration strengthened PSTs' instructional practice. They combined innovative ideas and used feedback to revise lesson content and pedagogical strategies while strengthening instructional practice. Collaboration exposed PSTs to new ideas to construct purposeful and engaging lessons. PSTs noted increased confidence, practical instructional use, and insight into how to adapt lessons and guide instruction for student learning. Because each member of the Jigsaw Lesson Study team taught the lesson, PSTs indicated they valued the observations, feedback, and revision processes. They noted that the modifications and instructional growth exhibited by the lesson revisions moved to clearer and more coherent instruction as the teaching progressed.

Study Guides Planning

PSTs valued the *study* step before planning. On their surveys, they stated that it guided their planning, strengthened content knowledge, and provided an opportunity to connect with their background knowledge. PSTs' responses revealed the *study* step promoted reflection on their past and present education. They reflected on their own positive and negative educational experiences that informed their research and planning decisions to craft meaningful and authentic learning opportunities for students. Additionally, *study* helped PSTs navigate the implementation of their team's ideas to collaboratively plan a lesson based on the difficult topics chosen by each team.

Observational Notes Strengthen Planning

According to the surveys, PSTs stated the observational notes were beneficial to reflect on student learning while a member of the team taught the lesson. Furthermore, PSTs noted how observational notes focused their lesson revisions and tracked learners' thinking and feedback. Many PSTs' reflections focused on how it was beneficial to observe students' engagement and reactions. PSTs had an increased understanding of student learning and they used this information to revise the lessons and create student-centered learning activities. Observational notes challenged PSTs to aim their attention on the learners of the lessons, finding importance even in the body language of the students. Based on PSTs' responses, the observational notes positively impacted PSTs' revisions of the lessons while considering student needs and interests. PSTs discussed how observing the teaching of lessons and providing feedback to the instruction impacted their thinking around planning. Reflections spotlighted how observing guided revisions to the lessons.

Takeaways from Jigsaw Lesson Study

It was evident from the PSTs that the collaborative and iterative nature of Jigsaw Lesson Study yielded a large variety of ideas that they believed helped with constructing and revising the lesson. PSTs also expressed how lesson study, specifically research and planning, helped them see alternative pathways for learning and planning. Furthermore, PSTs mentioned learning new instructional strategies and technological resources through Jigsaw Lesson Study. They realized revisions and feedback influenced planning and crafting an engaging and accessible lesson. PSTs noted how the process opened opportunities to think critically about lesson planning while considering the learners of the lesson and their thinking.

Concluding Thoughts

Based on the findings, we note that Jigsaw Lesson Study shares much in common with what other researchers of lesson study are finding. Similar to Ogegbo et al.'s (2019) findings, this kind of collaboration strengthened professional knowledge that included both instructional strategies and content. The findings here also lend evidence to Matney and Fox's (2022) conclusion that PSTs found it easier to focus their reflections on instruction to improve student learning, when doing so collaboratively with peers. Furthermore, as previously demonstrated by Cajkler and Wood (2016), PSTs grew as collaborative educators and learners, and developed a heightened awareness of how students learn. PSTs believed collaboration made their lessons better and gained professional learning because of the ideas that were shared to create the lessons, incorporating innovative strategies that were student-centered (Ogegbo et al., 2019). PSTs also learned that lessons may not always go according to plan, but because they are working together, they had the confidence to be flexible and modify their lessons, aligning with the research of Çevik and Müldür (2021). The PSTs found importance in researching, providing, and responding to feedback, and creating and teaching lessons to others. They found the Jigsaw Lesson Study experience to be well rounded and showed PSTs the importance of revising lessons and attending to different students.

Jigsaw Lesson Study can be implemented in teacher education courses in any content area and provides pedagogical opportunities for PSTs to grow as instructors while collaborating with their peers. By engaging in Jigsaw Lesson Study, PSTs are given an opportunity to understand each step within lesson study and PSTs will gain more confidence engaging in lesson study when they enter the field with in-service teachers. By giving PSTs knowledge of the professional processes involved in lesson study, teacher educators strengthen the entire field with new teachers who understand the process and value the professional outcomes. Jigsaw Lesson Study follows each step of the lesson study process and continues to promote the professional conversation that evolves when discussing instruction and student learning.

As novices, PSTs continue to develop notions about content and pedagogy throughout the teacher education program. Therefore, there is value in having them engage in Jigsaw Lesson Study because it allows them to grow in both content and pedagogy simultaneously. During Jigsaw Lesson Study, PSTs teach one another about the content and discuss various ways of teaching that content to strengthen student learning. The iterative use of lesson study opens opportunities for dialog about content and pedagogy through reflections and feedback. Even though PSTs are not teaching K-12 students, the PSTs learn content and pedagogical ideas from the professional collaboration with one another and their teacher educator. Although this would be enough to convince us to use Jigsaw Lesson Study with our PSTs before going into the field, there is another aspect of PST learning we have found to be important. They are learning how to professionally engage with one another in honest yet tactful ways, for the improvement of instruction. Hence, we find a triad of helpful aspects from enacting Jigsaw Lesson Study with our PSTs: Improved content knowledge, improved pedagogical knowledge related to that content, and knowledge of authentic professionalism in the field of teaching.

Teacher educators have a variety of contexts and challenges. One of those is taking people with an interest in teaching and turning that interest into professional practice. We have found that by providing iterative collaborative teaching experiences like Jigsaw Lesson Study, our PSTs experience growth, both personally as a teacher and professionally. A new awareness emerges that teachers do not need to exist in isolated classrooms, but through professional processes like lesson study, we can aid one another in overcoming our most difficult challenges.

References

Bjuland, R., & Helgevold, N. (2018). Dialogic processes that enable student teachers' learning about pupil learning in mentoring conversations in a lesson study field practice. *Teaching and Teacher Education, 70*, 246–254. https://doi.org/10.1016/j.tate.2017.11.026

Cajkler, W., & Wood, P. (2016). Adapting "lesson study" to investigate classroom pedagogy in initial teacher education: What student-teachers think. *Cambridge Journal of Education, 46*(1), 1–18. http://dx.doi.org/10.1080/0305764X.2015.1009363

Çevik, A., & Müldür, M. (2021). A model trial for the "teaching practice" course within Turkish teaching programs: Lesson study. *Journal of Language and Linguistic Studies, 17*(1), 403–422.

Erikson, F. (1986). Qualitative methods in research on teaching. In M. C. Wittrock (Ed.), *Handbook on research on teaching* (3rd ed., pp. 119–161). Macmillan.

Gillies, R. M. (2007). *Cooperative learning*. Sage Publication.

Glaser, B. G. (1978). *Theoretical sensitivity*. Sociology Press.

Glaser, B. G., & Strauss, A. L. (1967). *The discovery of grounded theory: Strategies for qualitative research*. Aldine de Gruyter.

Lewis, C. C., & Hurd, J. (2011). *Lesson study step by step: How teacher learning communities improve instruction*. Heinemann.

Matney, G., & Fox, M.. (2022). Examining programmatic lesson study in preservice teacher education. In S. Bateiha & G. Cobbs (Eds.), *Proceedings of the 49th Annual Meeting of the Research Council on Mathematics Learning* (pp. 86–94). Teaching and Learning Faculty Publications.

Ogegbo, A. A., Gaigher, E., & Salagaram, T. (2019). Benefits and challenges of lesson study: A case of teaching physical sciences in South Africa. *South African Journal of Education, 39*(1), 1–9. https://doi.org/10.15700/saje.v39n1a1680

Shuilleabhain, A. N., & Bjuland, R. (2019). Incorporating lesson study in ITE: Organizational structures to support student teacher learning. *Journal of Education for Teaching, 45*(4), 434–445. https://doi.org/10.1080/02607476.2019.1639262

Strauss, A. L. (1987). *Qualitative analysis for social scientists*. Cambridge University Press.

Sumarno, W. K. (2019). Investigating the impact of microteaching lesson study to the prospective English teachers' self-efficacy. *Journal SMART, 5*(1), 1–12. https://core.ac.uk/download/pdf/229584787.pdf

Weaver, J. C., Matney, G., Goedde, A. M., Nadler, J. R., & Patterson, N. (2021). Digital tools to promote remote lesson study. *International Journal of Learning and Lesson Study, 10*(2), 187–201. https://www.emerald.com/insight/2046-8253.htm

17

MATHEMATICS LESSON STUDY FOR EARLY PROFESSIONAL LEARNING WITH NOVEL PARTNERSHIPS

Christopher Nazelli and S. Asli Özgün-Koca

Teachers matter. The National Council of Teachers of Mathematics (2000) reminds us that

> students learn mathematics through experiences that teachers provide. Thus students' understanding of mathematics, their ability to use it to solve problems, and their confidence in, and dispositions toward, mathematics are all shaped by the teaching they encounter in school.
>
> *(p. 16)*

Unfortunately, there is evidence that many American students, especially those in large urban districts, do not experience high-quality instruction (National Center for Educational Statistics, n.d.). Researchers posit a connection between the quality of classroom instruction and the opportunities to learn that teachers encounter during their preservice preparation (Schmidt et al., 2011). In this chapter, we describe the essential features of two novel mathematics-focused lesson study activities designed to offer quality opportunities to learn for preservice teachers (PSTs) and present the PSTs' learning across four key goals of their early preparation.

Key Goals of Early Pre-Service Preparation

There is much to be done during the pre-service phase of teacher training. Feiman-Nemser (2001) described this as the period when PSTs must analyze their beliefs, increase their subject knowledge and knowledge of learners, build a beginning repertoire of teaching moves, and develop the tools to study teaching. These ideas have been echoed in the guidance and standards of key advocacy groups (Association of Mathematics Teacher Educators, 2017; Conference Board of the

DOI: 10.4324/9781003326434-22

Mathematical Sciences, 2012). We synthesize these as a call for PSTs to engage in activities that allow them to *construct a new vision of mathematics*, *create new knowledge for action, engage in the complexity of teaching,* and *prepare to learn in and from practice.*

Constructing a New Vision of Mathematics

For many PSTs, fluency with procedures and memorization were valued in their previous content courses and "doing mathematics in ways consistent with mathematical practice is likely to be a new, and perhaps, alien experience for many teachers" (Conference Board of the Mathematical Sciences, 2012, p. 11). Future teachers must experience mathematics as a dynamic field where they have opportunities "to construct viable arguments, to listen carefully to other people's reasoning, and to discuss and critique it" (Conference Board of the Mathematical Sciences, 2012, p. 33) so that they will be able to facilitate experiences for their future students. In short, PSTs must construct a new vision of what mathematics is and what it means to engage with mathematics.

Creating New Knowledge for Action

The National Research Council (2001) emphasized that "Effective programs of teacher preparation … cannot stop at simply engaging teachers in acquiring knowledge; they must challenge teachers to develop, apply, and analyze that knowledge in the context of their own classrooms so that knowledge and practice are integrated" (p. 376). In other words, teachers must *do* things with their knowledge. Rather than building inert knowledge that is separated from the practice of teaching, there is now an understanding of the power in building knowledge and immediately activating it (Darling-Hammond et al., 2009). This experience is especially useful when PSTs can observe highly proficient teachers engaging in the relational work of mathematics, as well as see students persevering, interacting, and exercising collective authority to validate each other's discoveries (McDiarmid, 1990).

Engaging in the Complexity of Teaching

Instead of allowing the urgency of practice into the teacher education curriculum, preparation programs typically break teaching down into fragments that can be practiced separately and then leave it to the PSTs to reconstitute the complex practice once in the classroom (Lewis, 2007). Both *creating new knowledge for action* and *engaging in complexity of teaching* reflect an emphasis on the situated nature of learning that would undergird a different type of preservice training (Lave & Wenger, 1991). It is important to note that "situated" here does not necessarily mean in a classroom full of children, but rather within authentic activities that serve the goals of the preparation program (Putnam & Borko, 2000). Exposure to authentic sources such as student work and video recordings can normalize the fourth component: Learning in and from practice (Ball & Cohen, 1999).

Preparing to Learn in and from Practice

Programs must prepare PSTs to be lifelong learners who draw from their practice, colleagues, curriculum, and research. Hiebert et al. (2007) offer a term for this experience: The PSTs must learn to take a "research stance" in their practice, engaging in "intentional learning from carefully planned experiences as part of the daily routine of practice" (p. 50). Taking a lesson as the unit of analysis, PSTs should be: (1) Guided to attend to the decisions involved in planning, in particular the learning goals (what the students should learn), (2) instructed on where to look for evidence of that learning, (3) asked to hypothesize about how the teaching decisions affected that learning, and (4) prompted to suggest improvements (Hiebert et al., 2007). This has also been described as "practitioner inquiry" (Hammerness & Darling-Hammond, 2005):

> The process of practitioner inquiry includes all aspects of a research or inquiry process: identifying questions of compelling interest … pursuing those questions through the collection of data (which may include observations of children, class or other observational field notes, or interviews with children, parents, or other teachers); and reflecting upon the questions through written work … and oral discussion with peers …
>
> *(p. 438)*

Lesson study offers a structure that helps PSTs adopt a beginning research stance that nurtures into full practitioner inquiry (Hammerness & Darling-Hammond, 2005) and, in addition, presents an environment rich in opportunities to learn to help meet the other three goals of early preservice preparation. With this in mind, we developed two mathematics-focused lesson study activities designed for PSTs, one in an early elementary content course and another in a secondary methods course.

Lesson Study Activities

In this section, we describe two lesson study activities for elementary and secondary education courses. These activities involved novel modifications that were driven by two factors: (1) The logistical conveniences of conducting an entirely (or almost entirely) intramural lesson study and (2) the lesson study participants being at an early stage of their pre-service education.

The Secondary Math Methods Course Lesson Study Activity

We designed an intramural lesson study activity, that involved two classes serving different PST cohorts within the same institution. The students in a secondary mathematics methods course (for clarity, we refer to them as "Secondary-PSTs") collectively produced a lesson for students in a content course for preservice elementary teachers (for clarity, we refer to them as "Elementary-PSTs"). *Studying*, *planning*, and *teaching* a lesson to the university students made this lesson study activity novel. The lesson study began with an expert teacher (one of the authors, a university faculty member) calling for a lesson on a specific topic. The Secondary-PSTs *studied and discussed* content and pedagogical resources on the topic of the lesson outside of the class time via discussion board. Approximately 20–25 minutes of each class session were devoted for lesson planning and occupied 8–10 weeks of a 15-week semester. The instructor of the content course visited the methods course during *planning* and provided feedback to fine-tune the lesson. The lesson was *taught* by the instructor and the Secondary-PSTs were invited to observe the live research lesson. The lesson was recorded so that, even if some could not attend, all Secondary-PSTs could observe successful and problematic areas of the lesson that they designed, as well as note any modifications made, on the fly, by the delivering instructor. During the debrief, Secondary PSTs and two instructors *reflected* on the lesson and its enactment. Finally, individual *reflection* was asked from Secondary PSTs on the whole process.

The Elementary Math Content Course Lesson Study Activity

We also designed a semester-long lesson study activity for a mathematics content course for Elementary-PSTs.[1] These Elementary-PSTs were in the first semester of their teacher preparation; thus, engaging them in lesson study activity at such an early stage made it novel. The activity occupied approximately 8 hours of class time, not including the research lesson observation or time spent outside class for individual reflections at key points in the cycle. The topic of understanding multiplication for third grade was provided by a team of teachers from a local elementary school with whom the authors worked in previous lesson study cycles as part of a long-term professional development series. The Elementary-PSTs *studied* the teacher team's district-mandated curriculum materials, practitioner research, and the content course material. They expanded a small segment of an existing lesson from the school district's mandated curriculum involving the commutative property of multiplication into an hour-long lesson and *planned* by anticipating student responses and considering other representations of multiplication (e.g., arrays and number bonds). The research lesson was *taught* by the content course instructor (one of the authors) to a class of third-grade students at the teacher team's school. All 11 Elementary-PSTs in the class, together with members of the teacher team, observed the lesson and *reflected* during a post-lesson discussion as well as individually in writing.

Modifications

From the descriptions above, significant modifications to the typical lesson study cycle are apparent in both the secondary and elementary models. The choice of lesson topic and pedagogical focus were controlled by the teacher educators and the teacher team, respectively; and this constitutes a significant loss of participant authority and autonomy inherent in K-12 lesson study practice. However, the early pre-service context required this adaptation as the PSTs were at the beginning of their professional learning for teaching. For the Elementary-PSTs especially, this course was the first in their preparation program; for some, it was their first course at the university. The lack of PST curricular knowledge prevented them from making informed decisions regarding the appropriate topic of the lesson to address student/school/district needs. In addition, both lessons were taught by an expert teacher rather than a lesson study PST participant. The lesson plans for both groups called for ambitious instruction involving high-leverage practices such as the connection of multiple representations and the facilitation of equitable whole-class discussions. The PSTs, although beginning to understand and appreciate the complexity of such instruction, were not yet ready to enact it. This modification allowed the PSTs to concentrate on the *study* and *plan* steps of the lesson study cycle without the added anxiety of having to deliver the lesson, and to focus entirely on student thinking during the *teaching* of the lesson. In the next section, we will see that, even with these necessary modifications, the learning of the PSTs was substantial.

Learning within the Lesson Study Activities

We organize the discussion of PST learning using our Framework of the Key Goals for Developing Mathematics Teachers (Figure 17.1). Within each goal, we present evidence from discussion transcripts and written reflections from both groups. As mentioned above, PSTs will be distinguished as Elementary-PSTs and Secondary-PSTs. All PST names are pseudonyms.

Construct a New Vision of Mathematics

The new vision of mathematics that the PSTs constructed during the lesson study activities encompassed both content and pedagogical aspects. Experiencing the

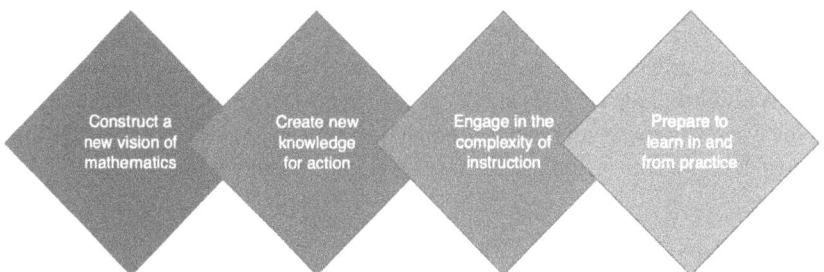

FIGURE 17.1   Framework of the Key Goals for Developing Mathematics Teachers.

difference between knowing mathematics for oneself and planning to teach others changed the PSTs' understanding of that content. For instance, Secondary-PSTs knew how to divide, and they might have used a model/strategy that made sense to them when other models might not. But when *studying* and *planning* for it, they needed to reflect on various models to help others.

> First, before this lesson study, I had no idea there were three specific division models. I realized while researching them that I have been using the missing factor method my entire life. It is important that I know all three models because different tasks require different models, and I need to be able to explain that to my students. Furthermore, the missing factor method won't make sense to a student who does not yet know their multiplication facts. For example, one of the students in my student teaching classroom was having trouble with division. I knew he had not memorized his facts, so I showed him the repeated subtraction method and he was able to evaluate the division expression. Studying division throughout this lesson study has helped grow my math content knowledge.
>
> *(Rebecca, a Secondary-PST, Reflection Paper)*

Here Rebecca not only reflected on how her content knowledge changed but also how that new content knowledge immediately affected her interaction with a high school student. Thinking about a mathematics topic with a student in mind changes how one views the topic itself. This transition from the doer of mathematics to the teacher of mathematics requires unpacking the content with a new vision. The lesson study provided multiple opportunities to learn in this regard, especially at the *study* step. Julia, an Elementary-PST, noted how the curriculum guidance on learning the multiplication facts for seven included a flexible method for computing 3 times 7,

> …what I wanted to highlight was how they did the sevens. So, if you're trying to get 21, they use the method, $14 + 6$ plus 1. When I was younger, they just told us to memorize the sevens … have us write it down. Just kept writing it down versus giving us different methods or different ways of solving the problem.
>
> *(Julia, In-Class Discussion Transcript)*

The *study* step allowed Julia to see mathematics differently, as a sense-making activity with multiple solution paths rather than memorization. This new vision of mathematics would take form for other PSTs in other steps as well.

Create New Knowledge for Action

For the new vision of mathematics to be operationalized, PSTs must create knowledge for teaching. The *study* step provided rich opportunities for the Elementary-PSTs to create new knowledge for action. When reviewing the suggestions of the curriculum guide, Jude connected the flexible thinking addressed in the document with our course work on addition algorithms and seemed to appreciate this new knowledge as knowledge *for action.*

> It looks like how they [the curriculum designers] broke it up was like in the ones place and the tens place. Kind of like how we practiced our [alternate algorithms for] addition. So that kind of made it easier for me to understand why they would reinforce that so much because *that helped us understand the actual process of addition* (emphasis added).
>
> *(Jude, In-Class Discussion Transcript)*

Jude took the knowledge that he constructed from our course activity, noticed it embedded in the curriculum guidance, and assessed its value in terms of teaching. The idea of student-centered teaching also became more defined and distinguished after observing the *teaching* of the research lesson for Justin, a Secondary-PST.

> Overall, the lesson study project was an eye-opening experience, and it was great to work with my fellow future educators. The lesson that was presented in Dr. Nazelli's class went well with students [Elementary-PSTs] collaborating and using the manipulatives to solve the problems given…When I have my own classroom, student engagement and multiple forms of representation is going to be my goal. Working in pairs or small groups can benefit the engagement in the classroom and having the students express their work in multiple ways can show how they learn.
>
> *(Justin, Reflection Paper)*

The Elementary-PSTs continued to comment on the value and emphasis of multiple solution paths, but now this idea was discussed in terms of *teaching.* Abbey, who earlier in the semester was concerned with how to reconcile a common curriculum and a student's thinking, was able to connect her new vision and knowledge of mathematics to teaching. Abbey credited her work during the intervening weeks and her examination of the curriculum materials enabled this shift.

> I know that it seems like it would be more challenging for the teacher, but I feel like it kind of makes it less daunting … because … *you don't have to make sure that everyone is on the exact same page.* In one of the sentences it says, "Apply properties of operations as strategies to multiply and divide." So, you present all this information, and the child chooses the one that connects with them and the way they conceptualize it … rather than telling them which way they conceptualize it with. So, I feel like it makes it almost easier for the teacher because you present them with the information and *they can understand it in their own way rather than saying, "You have to understand it this way."*
>
> *(emphasis added) (Abbey, In-Class Discussion Transcript)*

In Abbey's comment about what exploring multiple solution paths offers to students and teachers, we see a shift in her understanding of how a common curriculum can coexist with, and even support, students' individual thinking. We theorize that this shift was made possible by experiencing the diversity of thinking—and the valuing of that thinking—in our college course, and by connecting that experience to teaching practice through the lesson study's *plan, teach,* and *reflect* steps. This was also visible in the secondary methods course: "Differing ideas and discussions about how we should teach the lesson showed that there are multiple ways to teach the same thing and that each way might have a different purpose" (Anna, Reflection Paper). Anna began to understand the implications of this new vision of math class, where the teacher would have to anticipate and react to student thinking, and how she would now need to learn mathematics, for action.

Engage in the Complexity of Instruction

Experiencing the full act of teaching: Planning and then enacting the lesson provided multiple opportunities to learn from lesson study: (i) *Planning* collaboratively, (ii) observing the *teaching* of the research lesson, and (iii) *reflecting* on the research lesson. PSTs shared that, during the *plan* step, hearing the thinking of others illuminated the variety of options, each with affordances and limitations, to provide a higher cognitive challenge for the students. "Doing the lesson plan all together as a class I got to see everyone's idea about each section that I would have never thought of" (Valeria, a Secondary-PST, Reflection Paper). Additionally, the importance of planning in detail was brought up by Secondary-PSTs after *observing* the research lesson. For example, Anna, who previously advocated for supporting multiple ways of thinking, noted that

> I also learned through this project why it is important to anticipate every detail about the lesson including student responses and how to avoid misconceptions. I mentioned previously how anticipating responses could have improved the lesson in certain areas.
>
> *(Anna, Reflection Paper)*

Anna now highlights the anticipation of those potential various responses as an essential part of planning in detail. One Elementary-PST, Pam, was able to *observe* and *reflect* on the complex interplay of confidence, competence, and validation that took place during the lesson.

> For the tables that I watched, I felt like it was a lot of them following each other's lead. There wasn't that discussion … more or less … Penny watched a lot and she, like I had mentioned before, she had arrays drawn … both of them … and then she drew the number bond. Then she saw that someone had suggested number bonds to be drawn on the board; and she erased all her stuff. And [she]

left the number bond. Then she rewrote the arrays again; both of them again. And then somebody else put a number bond on the board and she erased them again! She kept going back and forth with it, but everybody at her table only had number bonds. She's also very shy.

(Pam, In-Class Discussion Transcript)

Jude, an Elementary-PST, noticed this as well, adding,

Yes, I was going to say … she looked at Mary's a lot, but Mary wasn't right. Penny was right, but she kept erasing her stuff … Because I feel like I was a lot like Penny when I was a kid. I probably knew the answer, but I was too scared to be wrong to vocalize it. But I feel like she was even more comfortable sharing her arrays after she got that *validation* (emphasis added).

(Jude, In-Class Discussion Transcript)

Thus, engaging in the complexity of instruction throughout the lesson study steps allowed the PSTs to appreciate and observe students struggling with mathematical content, practice, and identity.

Prepare to Learn in and from Practice

The lesson study as a whole provided an opportunity to learn. Still, the observation of the *delivery* of the research lesson that followed the *study* and *plan* step was particularly eye-opening for many PSTs.

I think having background study done before the lesson over the content *made the misconceptions more noticeable* (emphasis added). In class, we went over all the different ideas that the [elementary] students may have when asked to represent the story problem. Many of them clung to the number bond, which genuinely surprised me. We had all assumed the majority of representations would be arrays or tape diagrams.

(Pam, Research Lesson Reflection)

Here, the *study* and *plan* steps enabled Pam to *observe* student thinking during the lesson more deliberately especially in terms of mathematical representations. For Jude, the *study* step enabled him to anticipate student thinking and to *observe* how instructors can use these responses to orchestrate mathematical discussions.

My work on the curriculum had me looking for the thought process of students understanding the first story problem (emphasis added). It was interesting to see how many used the number bond and after looking around the class, I assume that is something they previously learned and seem comfortable with. It was super interesting to see how the teacher used guided questions and selected student examples to bring them all to the correct answer. It honestly shocked me how smoothly the commutative property fell into the lesson toward the end. Students exclaimed that 6×3 was the same as 3×6 even before the array was introduced. I was shocked at how accurate the dialogue provided by [the curriculum materials] played out into the actual lesson by the teacher.

(Jude, Research Lesson Reflection)

Within the brief reflections of Pam and Jude, we see a nascent "research stance" (Hiebert et al., 2007) which allowed them to make predictions about student thinking that were tested within teaching practice. This stance allowed Pam to be "surprised" and Jude "shocked" by what they saw. Secondary-PSTs also valued the whole process of lesson study as future teachers. They appreciated the collaborative aspect: "From start to finish the lesson plan study allowed us, as future teachers, to learn a useful process that we may use, as an individual teacher or as a collaborative mathematics cohort, to build our daily lesson plans" (Krista, Reflection Paper).

Interconnected Goals

Although we have presented the four goals separately, we must acknowledge the overlaps and interconnections among them. For example, the in-person *observation* of the lesson that occurred allowed Tiffany, a Secondary-PST, to see and appreciate her new vision of mathematics and teaching in action.

Being able to walk around the classroom and see people [Elementary-PSTs] working together and seeing how they show their work … Gaining more than an amplitude of knowledge to help me plan my lessons. As well as being able to work with the students more than just telling them the answer but using questions to help them work through the problem.

(Tiffany, Reflection Paper)

This was not what they experienced as doers of mathematics. Even though traditional methods "worked" for them, acknowledging that they might not work for all students was a crucial step as they were becoming teachers of mathematics.

Throughout my education, classes were for the most part traditional teacher-centered, especially with all my math classes. Since I liked and understood math this way, I never had a problem with it. But I also know that students learn differently from one another and that is why student-based learning is very important to incorporate in a classroom as a future educator.

(Valeria, a Secondary-PST, Reflection Paper)

Concluding Thoughts

The modified lesson study activities provided quality learning opportunities, and we believe that the evidence presented here indicates that the PSTs took advantage of them. These projects have shown us that novel partnerships, such as those between methods and content courses and between the university and K-12 settings, can be leveraged in creative ways to open up the possibility for highly productive lesson study, even at the earliest stages of preservice preparation.

Logistical and relational issues enabled and constrained these two modified lesson study activities. For example, the two university courses needed to be scheduled during the same semester; and we needed access to a K-12 classroom to deliver the lesson designed by the Elementary-PSTs. This was manageable, but collaboration between the Elementary-PSTs and the in-service teacher team was minimal due to scheduling constraints. The intramural nature of the Secondary-PSTs' lesson study meant that it was delivered to other PSTs, who had different learning goals than secondary students. However, the philosophical and pedagogical alignment of the two university educators, their classrooms, and the curriculum materials enabled the university courses' content and practices to serve as additional layers of curriculum study for both groups. It was often challenging to maintain equitable participation and contribution during the curriculum *study* and *planning* steps, as the groups were larger than a typical lesson study team. Our experience with lesson study facilitation as well as our use of shared document technology and group discussion protocols helped ensure that space was created for the ideas of all PSTs. These elements should be considered before attempting to implement such activities in other preservice contexts. Still, the substantial PST learning that can result from modified lesson study makes it an effort worthy of consideration.

Note

1 The 2019 Elementary-PST cohort was involved in both lesson study projects.

References

Association of Mathematics Teacher Educators. (2017). *Standards for Preparing Teachers of Mathematics*. http://amte.net/standards

Ball, D. L., & Cohen, D. (1999). Developing practice, developing practitioners: Toward a practice-based theory of professional education. In G. Sykes, & L. Darling-Hammond (Eds.), *Teaching as the learning profession: Handbook of policy and practice* (pp. 3–32). Jossey Bass.

Conference Board of the Mathematical Sciences. (2012). *The mathematical education of teachers II*. American Mathematical Society.

Darling-Hammond, L., Wei, R. C., & Johnson, C. M. (2009). Teacher preparation and teacher learning: A changing policy landscape. In G. Sykes, B. Schneider, & D. N. Plank (Eds.), *Handbook of education policy research* (pp. 613–636). Routledge.

Feiman-Nemser, S. (2001). From preparation to practice: Designing a continuum to strengthen and sustain teaching. *Teachers College Record (1970)*, *103*(6), 1013–1055.

Hammerness, K., & Darling-Hammond, L. (2005). The design of teacher education programs. In L. Darling-Hammond, & J. Bransford (Eds.), *Preparing teachers for a changing world: What teachers should learn and be able to do* (pp. 390–441). Jossey-Bass.

Hiebert, J., Morris, A. K., Berk, D., & Jansen, A. (2007). Preparing teachers to learn from teaching. *Journal of Teacher Education*, *58*(1), 47–61.

Lave, J., & Wenger, E. (1991). *Situated learning: Legitimate peripheral participation*. Cambridge University Press.

Lewis, J. M. (2007). *Teaching as invisible work*. (Unpublished doctoral dissertation). University of Michigan, Ann Arbor, MI.

McDiarmid, G. W. (1990). Challenging prospective teachers' beliefs during early field experience: A quixotic undertaking? *Journal of Teacher Education*, *41*(3), 12–20.

National Center for Educational Statistics. (n.d.). *National assessment of educational progress*. https://nces.ed.gov/nationsreportcard/

National Council of Teachers of Mathematics. (2000). Principles and standards for school mathematics. National Council of Teachers of Mathematics.

National Research Council. (2001). *Adding it up: Helping children learn mathematics*. National Academy Press.

Putnam, R. T., & Borko, H. (2000). What do new views of knowledge and thinking have to say about research on teacher learning? *Educational Researcher*, *29*(1), 4–15. 10.2307/1176586.

Schmidt, W. H., Cogan, L., & Houang, R. (2011). The role of opportunity to learn in teacher preparation: An international context. *Journal of Teacher Education*, *62*(2), 138–153.

CONCLUSION

Finding Connections through Form

Sharon Dotger, Gabriel Matney, Jennifer Heckathorn,
Kelly Chandler-Olcott, and Miranda Fox

In 2008, Sharon traveled to San Francisco, California, to attend a meeting for teacher educators with federal grant funding to collaborate with mathematics and science public school teachers. While she was aware of lesson study from reading *The Teaching Gap* (Stigler & Hiebert, 1999), a book that summarizes the findings from the Third International Mathematics and Science Study, she was not aware that lesson study was happening in the United States until Catherine Lewis took to the podium and overviewed her team's local work. Inspired by Catherine's explanation and descriptions of other lesson study projects in the United States, Sharon attended a lesson study conference organized by the Chicago Lesson Study Alliance.

During the post-lesson discussion of the live research lesson at the conference in Chicago, she heard international mathematics education scholar Ban Har Yeap discuss the concept *shu-ha-ri*. *Shu-ha-ri* loosely translates as "learn the form, master the form, depart from the form" and is often attributed to Japanese martial arts. In martial arts, "forms" are choreographed series of blocks, kicks, strikes, and stances that help the practitioner build muscle memory and automaticity and study the flow between one movement and the next. Each form has a specific set of movements, done in a particular order, and the forms get progressively more complex as the practitioner advances in their study of the martial art.

In the lesson study context, *shu*—learn the form—can be thought of as learning the steps of lesson study. This might initially manifest as learning the what and the how of the *prepare*, *study*, *plan*, *teach*, and *reflect* steps. Learning these steps means understanding the context within which they are used, and the first section of this book aims to illustrate that context with a broader view in order to help situate each step within the larger trajectory of lesson study. Key takeaways from Section I included the knowledge needed to understand the relationship between lesson study

DOI: 10.4324/9781003326434-23

and preservice teacher education and how the teacher preparation structures in the United States lend themselves to adopting and adapting lesson study. In Chapter 1, Heckathorn and Dotger explained how the lack of a shared practice or programming in teacher preparation contributes to variance in lesson study implementation with PSTs. In Chapter 2, Kennedy and Wilcox showed readers how lesson study helps PSTs understand how their decision-making relates to student learning. Then, in Chapter 3, González and colleagues introduced the key roles that teacher educators and/or cooperating teachers must play when facilitating lesson study, focusing on how each role requires a distinct skill set and arguing that the facilitator must know how to move fluidly between those roles.

Learning about the form of lesson study, more *shu*, continued in Section II. Each chapter in Section II provided nuance about the steps of lesson study. In Chapter 4, Hummer and Lesseig highlighted the structures and tasks that are necessary to *prepare* for a strong lesson study cycle for PSTs. And in doing so, they made visible some of the hidden labor of teacher educators. Rogers and colleagues, in Chapter 5, urged fellow teacher educators to slow down their lesson study cycles to allow PSTs the time needed to adequately *study* curriculum and instructional materials before commencing the plan step. Then, in Chapter 6, Reins and Melville advocated for enculturating PSTs into good teaching through careful and purposeful *planning* that sets up a strong lesson and, more importantly, provides the collective PST group with the capacity to learn from the teaching of the lesson. In Chapter 7, Michaels and Glen provided readers with examples of the multiple ways to teach and observe a research lesson when lesson study is practiced with PSTs. Finally, in Chapter 8, Dotger and Chandler-Olcott illustrated a variety of ways the *reflect* step can be incorporated into lesson study. Taken separately, each of these chapters provided input into one step of the lesson study cycle. However, we contend that the greatest amount of learning is generated by lesson study cycles that incorporate all five steps, pointing to how learning is carried forward from one step to the next throughout the cycle.

The chapters in Section III showed readers how lesson study can be used with PSTs to learn more than just content or instructional practices. It can assist them in learning about and implementing core ideas around diversity, equity, inclusion, and justice (DEIJ) too. In Chapter 9, Graham and Roth McDuffie described their own work using lesson study to engage PSTs in DEIJ. They learned that although PSTs can notice these ideas, in order for them to plan meaningfully for them, teacher educators must deliberately ask PSTs to attend to DEIJ. Then Taylor and colleagues, in Chapter 10, pointed out how trust is a key component for communication and collaboration to serve anti-racist teaching agendas. Finally, in Chapter 11, Kalinec-Craig and colleagues framed DEIJ learning through the pairing of the Torres' Rights of the Learners framework with lesson study, pointing out, as others have throughout the book, how lesson study can be partnered with multiple frameworks (e.g., TRU, noticing, Clough's Decision-making Framework, and Torres' Rights of the Learner) to encourage PSTs to advance equity in their instruction.

In Section IV of this book, examples of the adaptability of lesson study were on full display. Johnson and colleagues, in Chapter 12, demonstrated how lesson study with PSTs can set a foundation for collaboration that can continue in their in-service practice. In Chapter 13, Hagevik and Falls addressed adapting lesson study to an online environment and urged readers to simultaneously attend to the core ideas of lesson study and principles of high-quality online teaching. Roehrig and Suh, in Chapter 14, pointed out that the quality of PST's reflection depends on what teacher educators help them notice during observation of instruction. Likewise, in Chapter 15, Haydar and colleagues pointed to how lesson study can be used to advance PSTs' knowledge for working across content areas to advance interdisciplinary connections. Then, in Chapter 16, Weaver and Matney introduce the idea of jigsawing lesson study to maximize PSTs' opportunities to both teach and observe. Finally, in Chapter 17, Nazelli and Ozgun-Koca demonstrated how novel partnerships can be leveraged in creative ways to open the possibility for productive lesson study with PSTs.

It is *ha*—master the form—that we believe is marked across the chapters in this book. Throughout the book, the authors have demonstrated how lesson study can be done—and how it can be stretched based on context and situations that provide boundaries for the teacher educators' abilities to implement a full cycle. But the authors also do two other key things—first, they show how lesson study can provide a framework from which teaching and learning can be studied and improved. This book is full of examples of teacher educators using lesson study to understand better a specific part of their practice or PSTs' learning. Second, they show us that lesson study, no matter how long one has been involved in it, is never really "mastered." Veteran teacher educators with longstanding lesson study practice are still learning, for example, how to be helpful knowledgeable others, how to design lesson study cycles that emphasize collaboration between host teachers and PSTs, and how to build *study* steps that engage PSTs in building equity in their own lessons. Thus, "ha" becomes the point whereby the lesson study practitioner, be it the teacher educator, the preservice teacher, or the in-service teacher, can study the why and when of each step, the reasons that undergird each step, the ways the steps are connected to one another, and the timing for transitioning from one to the next and back again. Furthermore, we think this is likely the time when the less visible benefits of lesson study, such as changes in knowledge, beliefs, practices, agency, and collegial social capital (Lewis, 2020), begin to be experienced and noticed by practitioners.

This book also embraced *ri*—departure from the form—of traditional, full cycle lesson study. *Ri* is where innovation emerges, likely in the subtlety of the practice, such that other users of lesson study can recognize both its similarity to the fundamentals and the artistry of its expression.

A core learning for us, as we edited this book, revolved around teacher educator learning. We assumed that most teacher educators deepened their knowledge of content, teacher learning, and constructs like pedagogical content knowledge

(PCK) and its variants during their doctoral program. Through this preparation, they began to learn the form of teacher education. As they continue into their careers, they built from this preparation and their years of prior experience in school classrooms. From these sources of study and experience, they moved from *shu* to *ha* in teacher education, continuing into *ri* over time. Yet, few have opportunities to learn about lesson study along the way. Thus, although teacher educators have developed expertise in certain aspects of teacher education that overlap with some aspects of lesson study, learning about lesson study places them back in a place of learning a new *shu*. They are, in essence, learning a new form of teacher education.

This process applies, perhaps most visibly, to the role of knowledgeable other. As "teachers of teachers," teacher educators in the United States have worked for over 100 years on developing methods for preparing teachers for the complexities of leading the learning of their students. Yet, across the book, multiple authors point out the variations in facilitation, questioning, feedback, structures, and tools that teacher educators are using across the lesson study cycle.

Teacher educators, as traditionally prepared, are not automatically knowledge-able others for lesson study. While they certainly have knowledge and skill relevant to lesson study, we argue they need to become students of lesson study, just as they studied to become a teacher educator in the first place. Like classroom teachers, teacher educators make myriad choices as they work with teams of PSTs across the lesson study process. For example, during the post-lesson discussion, if teacher educators are working alone, they must decide when or if to shift the conversation away from a description of what the students thought during the lesson and how and toward a discussion of the implications of those observations for future instruction—within the immediate unit arc and beyond. We believe more can be learned by studying that transition, as well as the many others that teacher educa-tors make in other areas of the cycle.

Throughout the book—just like in martial arts—learning the nuances of lesson study remains key. To connect back to the material arts framing, white belts, the newest practitioners of an art, learn front stance. Across the decades of practice it would take to reach the highest levels of the art, front stance remains. However, subtleties become evident over time: The turn of the toes, the balance of weight side-to-side and front-to-back, the angle created by the line down the spine to the hip. In the west, lesson study practitioners, even those of us with the most experi-ence, have only had 20 years or so to learn this craft. Naming the nuances and sub-tleties of lesson study is difficult, especially as variance in lesson study abounds. But finding the descriptions of the nuances is possible. As we continue to work on lesson study as a community, we can continue to study our fundamentals to illumi-nate the features of lesson study that are not apparent at first glance.

As we work together to study the fundamentals of lesson study and find the form with PSTs, we are struck, yet again, by the sense of connection that is embedded within it. Lesson study, really, is a process of coming to connect. Through lesson study, teacher educators connect with each other to discuss how lesson study can be used to promote PST learning and to invite each other into our classroom spaces as knowledgeable others. We also connect with our students—drawing on the trust we create in the classroom to study lessons. We connect over and with content, attempting to improve our pedagogical core, and we connect our students to one another and the content. Finally, we connect our PSTs to schools, host teachers, and K-12 students. In fact, we noticed in the Preface that Catherine Lewis noticed new connections within the book that augmented our organization, further evidence of the possibilities of connection building through lesson study.

In recent work, Lewis (2020) reminds us that lesson study is not only about the steps. If we focus too much on them, we lose track of the underlying benefits and possibilities of doing lesson study. We devoted time to the steps in this book because they help us break down the complexities of a complete lesson study cycle and specify how to do it. Essentially, they help us articulate the *shu*. Yet, isolated, they are less meaningful. As book editors, we have learned that we need to articulate more clearly the significance of the connectivity between the steps, the way one aspect of the cycle feeds into the next. We see this connectivity within the in-service context, because when lesson study is a school-wide practice, lesson study cycles feed into one another. Over time, these cycles lead to changes in instructional practice and a more shared vision of high-quality instruction among the faculty. There are also changes in the way teachers view and use instructional materials. In fact, in Japan, the accretion of findings from teachers' lesson study work informs the adjustments to their national standards and their instructional materials. These lessons from the in-service context suggest that a turn to emphasizing connectivity in lesson study is likely a beneficial direction to take for teacher educators, PSTs, host teachers, and public-school learners.

We set forth to hold the Lesson Study for Mathematics and Science Teachers Educators Conference and write this book to find other teacher educators engaging in lesson study practice with their preservice teachers. We wanted to make our current practices visible to ourselves and others to facilitate more discussion, more refinement of practice, and, frankly, more lesson study. The chapter authors in this book have shared with us their ideas, showing us where they are in their own journey of learning about and teaching others to use lesson study. As the editors of this book, we believe that as we continue in our own lesson study practice, we will continue to learn, to notice nuances better, and to decide how and when to reveal those nuances to novices in our teacher education work. Books generally have a way of suggesting definitive ideas, and their permanence may suggest finality. Instead, we view this book as a documentation of a moment in time, a place where we are today, shared so that others may join us in our study of how to prepare teachers in ways that create positive learning experiences for them and their future students. We are grateful to all the authors who shared their time and learning with us.

References

Lewis, C. (2020). Conclusion: How do we judge the success of lesson study adaptations? In *Stepping up lesson study* (pp. 116–120). Routledge.

Stigler, J. W., & Hiebert, J. J. (1999). *The teaching gap: Best ideas from the world's teachers for improving education in the classroom.* Simon and Schuster.

INDEX